Practical

Plant
(partially obscured)

es

ding
and
ility

GE

MECHANICAL ENGINEERING
A Series of Textbooks and Reference Books

Founding Editor

L. L. Faulkner

*Columbus Division, Battelle Memorial Institute
and Department of Mechanical Engineering
The Ohio State University
Columbus, Ohio*

Practical Plant Failure Analysis

A Guide to Understanding Machinery Deterioration and Improving Equipment Reliability

Neville W. Sachs, P.E.

Taylor & Francis
Taylor & Francis Group
Boca Raton   London   New York

CRC is an imprint of the Taylor & Francis Group,
an informa business

CRC Press
Taylor & Francis Group
6000 Broken Sound Parkway NW, Suite 300
Boca Raton, FL 33487-2742

© 2007 by Taylor & Francis Group, LLC
CRC Press is an imprint of Taylor & Francis Group, an Informa business

No claim to original U.S. Government works
Printed in the United States of America on acid-free paper
10 9 8 7 6 5 4 3 2 1

International Standard Book Number-10: 0-8493-3376-8 (Hardcover)
International Standard Book Number-13: 978-0-8493-3376-7 (Hardcover)

Visit the Taylor & Francis Web site at
http://www.taylorandfrancis.com

and the CRC Press Web site at
http://www.crcpress.com

Preface

The world of failure analysis began opening up to me about 30 years ago when, as a young engineer on a plant tour, I was shown a row of six 600-hp pumps that had broken 11 pump shafts in the previous 8 or so months. The pumps were impressive, but the fact that they were repeatedly breaking 6-in.-diameter shafts was just amazing to me. After many months of charting and analyzing motor power requirements, pump flow rates, liquid temperatures, and inspecting everything else we could think of, we eventually realized that the pumps were operating far back on their curve and fatigue pulses occurred every time a vane passed the cut water. The result of this cyclical loading, combined with the shaft corrosion, was causing the failures.

In those days, there was a tremendous amount of material on mechanical design and failure analysis available in print. If the people who approved the changes that lead to the pumps running at that low flow rate had an understanding of the effects on the mechanical reliability, they probably would have made some wiser choices. If we had known what to look for we would have solved the problem much faster. Unfortunately, they didn't understand that the pumps really were not suited to the job ... and it took us several months to learn why they were failing.

Today, we have all that print information, plus the availability of the Internet but, from what we have seen while traveling around North America, the answers aren't easy to find, and many folks don't understand how to look for them.

What I've tried to do is write a book that engineers and skilled maintenance people can use as a practical to guide understanding how and why failures occur and how to prevent them from recurring. The first two chapters address some of the basics behind failure analysis. The next six chapters discuss materials and how to identify the common failure mechanisms. The last section explains how specific components, belts, gears, bearings, etc., work and how those basic failure mechanisms apply. In addition, they show how, with careful inspection, the multiple physical causes can easily be diagnosed.

While I was learning about failure analysis, several of the instructors mentioned that 90% of the physical roots of all failures could be discovered by a detailed inspection of the site and the parts, using good lighting and low-power magnification. This book should make it easier for the readers to do just that.

One of the things we have tried to do in the book is include a series of diagnostic charts that can be used to help guide the failure analysis procedure. (I say "we" because I had a tremendous amount of guidance and comments from other folks, especially the rest of the crew at Sachs, Salvaterra & Associates [SS&A], in developing these charts.) We realize that there is no true substitute for experience, but these charts should give analysts a great start in the right direction and enable them to gain that experience more rapidly.

Just as failures are always the result of multiple causes, this book is the result of multiple contributions to my education, and there are a lot of people who should be thanked for helping me along the way. Charles Latino, an industry pioneer, was the prime mover behind the reliability engineering effort at Allied Chemical without which there never would have been a job for me. (He founded the Reliability Center, and we continue to work with him and the entire Latino family.) Walt Shepard was my first supervisor at Allied Chemical, and he successfully encouraged me to pursue failure analysis and nondestructive inspection applications while letting me think they were my idea. Other folks such as Dick O'Brien, Darren Bittick, Mike Altmann, Tom Brown, Hans Nichols, Dan Carroll, Tony Kinsch, Rick Kalinauskas, and many others have helped by both supplying and helping to solve a fascinating variety of failures and reliability challenges that we have all learned from.

Another group that has played a tremendous support role is the rest of SS&A. We've been in business for over 20 years and Phil, Eddie, Steve, Dale, and Dwynn have been a tremendous resource (and have tried to keep me in the real world) and make work an enjoyable place.

Most important on the list of people to thank is a very patient wife, Carol Adamec, who put up with months of my partial withdrawal from civilization while writing the book.

One last thought. People ask me what portion of the failures we've investigated has involved human errors as a contributor. My response is that we have never seen a failure that didn't involve human error as a substantial contributor. All humans occasionally make mistakes, even those who write texts on failure analysis. If you find an error in the book, I would appreciate your input.

Neville W. Sachs, P.E.

The Author

A native of Northern New Jersey, Neville Sachs attended Stevens Institute of Technology in Hoboken, New Jersey. A member of the class of 1963, he received a Bachelor of Engineering degree, majoring in mechanical and chemical engineering.

After 10 years in a variety of manufacturing, engineering, and supervisory positions, he joined Allied Chemical (now Honeywell International). From then until the Syracuse works closed in 1986, he was deeply involved with maintenance and machinery reliability, first as an engineer, then as the supervisor of the Reliability Engineering Department. While with Allied, he was instrumental in developing one of the first large predictive maintenance inspection programs in the nation, served on a number of corporate technical committees, conducted numerous failure analyses, and received a patent for a device that demonstrates several of the mechanisms of fastener failures.

In early 1986, Sachs, a licensed professional engineer, joined with Philip Salvaterra to form Sachs, Salvaterra & Associates, Inc. (SS&A). This corporation, based in Syracuse, New York, provides technical support services generally directed toward improved plant and equipment reliability. Among their capabilities are detailed professional engineering studies, mechanical failure analysis, corrosion and materials engineering, vibration monitoring, root cause analysis, specialty tank, vessel and piping inspections, design reliability analysis, and a wide variety of nondestructive examination methods. SS&A also provides training in many of these areas.

The author has conducted hundreds of failure analyses and failure analysis classes across North America. He is a member of ASM International, where he was active on a national committee, and is past chairman of the Syracuse Chapter, ASME, the National Association of Corrosion Engineers, the Society of Tribologists and Lubrication Engineers, where he is active on a national committee, the NSPE, and the Technology Alliance of Central New York. He is a frequent speaker for both regional and national programs, has contributed significant sections to two books, and has written over 40 technical articles and papers for U.S. and European magazines and journals, primarily on failure analysis and equipment reliability.

He and his wife, Carol Adamec, an accomplished artist, thoroughly enjoy skiing, tandem bicycle touring, hiking, kayaking, and chasing after nine grandchildren.

To Lauren

Table of Contents

1 An Introduction to Failure Analysis

The people we've worked with began doing failure analysis on industrial equipment in the mid-1960s. Their efforts were primarily linked with an interest in improving equipment reliability and production capacity in chemical plants. Prior to that time, most industrial failure analyses merely involved trying to understand the physical roots of the problem. From their work and a manufacturing and processing viewpoint, it wasn't until the early 1970s that the realization began to develop that the true sources of our problems were much more complex.

What is *failure analysis*? There are probably as many definitions as people you ask the question of, but we prefer to think of it as "the process of interpreting the features of a deteriorated system or component to determine why it no longer performs the intended function." Failure analysis entails first using deductive logic to find the mechanical and human causes of the failure, and then using inductive logic to find the latent causes. In addition, it should also lead to the changes needed to prevent the recurrence of failure.

Many people in industry prefer not to use the term *failure analysis*, and we have heard the statement more than once that "We don't want our maintenance improvement (or some similar) program being driven by concentrating on failures." It's easy to understand their words but impossible to understand their logic. What better and more efficient way is there to improve than to intelligently look at our weaknesses and mistakes, learn from them, and try to do a better job the next time?

This book is my attempt at a manual that explains how and why mechanical machinery fails. To do that, I have tried to explain how the basic failure mechanisms occur, the things we all do to cause machinery problems, how to recognize those things, and what to do to prevent failures in the future.

THE CAUSES OF FAILURES

Why are there premature equipment failures? When the people involved are asked this question, the answer is almost always "the other guy's fault." If one were to ask a plant millwright or a maintenance mechanic, the most likely answer to that question would be "operator error." But if the same question were asked of an operator who worked in the plant with that millwright, his or her answer might be "it wasn't properly repaired." At times there is some validity to both these answers, but the honest and complete answer is much more complex.

It would be nice and neat if there were only one cause per failure, because correcting the problem would then be easy. However, in reality, there are multiple causes to every equipment failure. Unfortunately there are many people who believe that there is only one cause for a failure. However, look at any well-studied major disaster and ask if there was only one cause. Was there only one cause for Three Mile Island? The Exxon Valdez mess? Bhopal? Chernobyl? A major airplane crash? Looking at these and other well-recognized and extensively studied failures, reports show that they all have multiple causes. Then why would an intelligent person believe a typical pump or fan failure would be different? (In the case of Three Mile Island, there were three huge studies, each commissioned by one of the responsible groups. All three of the studies said there were numerous causes, but that it was "primarily the fault of the other organizations".)

1

At an international conference on failure analysis, a presentation was made on the causes of aircraft equipment failures. The data from the presentation is shown in the following:

- 30% — Manufacturing errors
- 26% — Design errors
- 23% — Maintenance errors
- 18% — Material selection
- 3% — Operation

During the question-and-answer session after the presentation, a member of the audience asked the speaker why they had listed only one cause for each failure when there were usually multiple causes. The speaker agreed with the questioner, then said " … there was only one blank on the form." This answer is a quote and an interesting testimony to the public's lack of perception. They think about cases that have been carefully studied, such as those listed earlier, and agree that there are obviously multiple causes for each. Yet, when directly involved with a problem, many people still ignore reality and come up with data such as that shown in the preceding study. They then take this data, draw an attractive pie chart or bar graph, and point to a nice neat single cause for every failure — when simple and honest logic clearly states that isn't true.

Two comments:

At a later time in the session, another group analyzed the same basic data set and reached very different conclusions. They too sorted the data with the idea of a single cause for each failure! One of the questions I find interesting is, "Why don't they recognize that many failures have more than one physical cause and all failures have more than one human cause?" I suspect that there is an innate human desire to neatly categorize problems and answers but the answer to this lies with people who are more skilled than I am.

Trevor Kletz has written a fascinating series of books detailing events leading to many of the world's major industrial disasters. In his approach to failures he uses a three-tiered system for classifying the causes; we shall use a similar method. In this approach there are three general classes of failure causes or roots, and, until the investigator understands each of these classes and their inevitable interaction, they do not truly understand how and why the failure occurred and don't have the ability to prevent another.

The divisions we have used for these roots and some examples are:

Physical roots — This is the physical mechanism that caused the failure, i.e., the shaft failed from rotating bending fatigue complicated by both corrosion and a significant stress concentration.
Human roots — This is how "inappropriate human intervention" resulted in the physical roots. Continuing the preceding example:
  When the shaft was designed, the engineer didn't expect the corrosive conditions and used an alloy that was relatively sensitive to corrosion.
  The machinist who made the shaft cut a sharp corner where there should have been a gentle radius.
  The millwright who installed the machine misaligned it significantly, causing vibration and an unanticipated bending load on the shaft.
Latent roots (system weaknesses) — These are the corporate policies or actions that allow "inappropriate human action." Again pursuing the previous example:

The engineer was pressured by supervision to rapidly complete this assignment and didn't carefully analyze the full range of conditions the machine was going to experience. In addition, he was only two years out of college as a mechanical engineer, didn't do well in his materials classes or really understand the use of different alloys, and his supervisor had no idea of these weaknesses.

After the machinist made the shaft, it was shipped to the plant for installation. Unfortunately, there was no receiving inspection for materials used in maintenance at the plant, and the shaft was accepted with the error.

The maintenance management had stated that the millwrights should be using laser alignment equipment, but the policy wasn't enforced, and this mechanic hadn't been to the alignment class. (It was a "hot job" and had to be back in operation ASAP, so they couldn't waste time on alignment.)

One truly significant industry dilemma is that problems that are not the result of catastrophic failures are frequently not recognized at all. Often, normal corrosion and wear problems, in which slow wastage destroys a component or a piece of equipment, are not recognized as opportunities for improvement and are ignored. An unexpected leak in a pipe or vessel may be recognized as a failure but the long-term deterioration of a tank or the routine replacement of a set of V-belts is thought of as being "just part of the cost of doing business."

Much of the routine maintenance that a plant or facility performs falls into this "ignored failure" category. During a recent assignment in a plant that was generally accepted in their industry as having an excellent maintenance program, one of their instrument technicians found that they were spending over $300 per day replacing gas system filters that should have cost less than a tenth of that. In a similar plant, extensive interviews with maintenance personnel found that almost 20% of the routine maintenance budget was the result of recurring items that should never have happened in the first place. In both situations, the problems had been in existence for years but had never been recognized.

ROOT CAUSE ANALYSIS (RCA) AND UNDERSTANDING THE ROOTS

The concept of RCA has many interpretations. We use it to get to the true causes of a failure but many folks like to stop at the physical causes. If you solve for just the physical causes of the failure, all you have done is to determine why that one incident occurred. Going into greater depth and finding the human roots allow one to change human behavior and eliminate a group of failures. But further reaching into and correcting the latent weaknesses result in whole classes of failures being eliminated. As an example of this, look at the change in automotive reliability after the introduction of Japanese imports in the 1970s.

PHYSICAL ROOTS

The *physical root* is the mechanism that caused the part to break. It may be fatigue, overload, wear, corrosion, or any combination of these. The importance of understanding the physical root or roots cannot be overstated. Failure analysis must start with accurately determining the physical roots, for without that knowledge, the actual human and latent roots cannot be detected and corrected.

Most of this book is about the physical roots of failures and there are times when the cause is a single physical root. However, our data from years of analysis shows that there frequently is more than one root.

Human Roots

The *human roots* are those human errors that result in the mechanisms that caused the physical failures. A good nonindustrial example of a human root of a problem would be an automobile driver's use of a cell phone and the effect of this on accident rates. Two studies on cell phone usage show:

> Motorists are four times more likely to crash when using cell phones (*New England Journal of Medicine*, 1997).
>
> Mobile phone use while driving hugely increased the chance of an accident. Increases, ranging from 34% to more than 300%, have been cited in different studies.*

Other studies have reported similar data. It seems fairly obvious that these drivers don't intentionally have collisions and that it is the cell phone use and the resultant distraction that result in both an increase in the rate of human errors and the increase in accident rates.

Good statistics on the frequency of human error in industry are generally difficult to find. However, an article in *Chemical Engineering* magazine stated the following incidences for a variety of human errors in industry.

Industrial Activities

- Critical routine task — 1/1000
- Noncritical routine task — 3/1000
- General error rate for high-stress, rapid activities — 1/4
- Nonroutine operations (startup, maintenance, etc.) — 1/100
- Checklist inspection — 1/10

General Human Error

- Of Observance — 1/50
- Of Omission — 1/100

There are texts on human error and the data shown in the following is by Swain and Guttman and provides an estimate of the frequency of human errors in both plant and everyday situations. (Parenthetical values indicate 5th and 95th percentiles.)

1. Select wrong control in a group of labeled identical controls: 0.003 (0.001 to 0.01)
2. Turn control in wrong direction in a high-stress situation where design is inconsistent: 0.5 (0.1 to 0.9)
3. Operate valves in incorrect sequence [less than 10 sequences]: 0.01 (0.001 to 0.05)
4. Failure to recognize an incorrect status when checking an item right in front of you: 0.01 (0.005 to 0.05)

The number of errors cited in the preceding text looks daunting. However, Charles Latino, president of the Reliability Center, Inc., Hopewell, VA, a noted human reliability expert, has stated that "Human error experts say that the average person makes about six significant errors per week." (A *significant error* is defined as an error that could result in significant economic loss or personal injury.) There have been studies of human error, some of which suggest that Mr. Latino's figures may even be conservative.

---

* *USA Today*, October 19, 2000 (from a 1999 report by the Center for Urban Transportation).

Trying to understand the reasons for these errors is an almost endless task. One significant point is that we, as humans, have an unrealistic opinion of our abilities and generally overstate them. In Tom Peters' work, *In Search of Excellence*, he describes surveys in which students are asked to evaluate their abilities. The self evaluations on "the ability to get along with others" and "leadership ability" seem grossly overestimated, but we don't have a way to evaluate these skills in a comparative manner. However, when 94% of these students rate themselves as having above-average athletic ability and 60% rate their athletic ability as being in the top 25%, we know that they (and, unfortunately, the rest of us) don't have a realistic assessment of their skills.

In an effort to understand more about this persistent tendency to overestimate our capabilities, we conducted a series of surveys of industrial and commercial workers with the following results:

1. In one particularly hazardous occupational group, in which 10% of the employees had a lost time injury every year, the average worker rated their safety awareness as being in the 83rd percentile.
2. In a similar group, the average worker felt they made a significant error about once every 8 months, i.e., less than 1% as often as the human error experts indicated.
3. In a third set of studies of several thousand workers and engineers in the process industry, the average worker indicated that their skills were in the 68th percentile level when compared with their peers, that they made a significant error about once every 5 months, and their peers made twice as many errors as they did.

Consistent with these studies, one of the things we have found in conducting failure analyses is that the typical (and frequently unstated) reaction to a failure is that "It must be the other guy's fault." Maintenance people really believe the problem is operator error. Operators really believe the maintenance people don't fix the machines properly, and it is not uncommon for engineers and supervisors to believe that everybody else is at fault. In an effort to better understand more about the causes of equipment failures, we reviewed 131 detailed failure analyses that our company conducted during a 3-year period in the late 1990s.

From the 131 analyses, the major physical failure mechanisms were (percentages do not total 100% because of rounding):

Types of Physical Failure	Number of Analyses	Percentage of Total
Corrosion	23	18
Fatigue	57	44
Wear	15	11
Corrosion fatigue	17	13
Overload	19	15

Two important points must be noted:

1. In defining these five categories, there is the possibility of confusion between corrosion fatigue and fatigue. Our practice was to assign fatigue as the primary mechanism to those cases in which the component would have eventually failed and corrosion was not needed to affect the failure. In those situations in which the component would not have failed without the action of the corrosion, i.e., there was cyclic loading but it was not severe enough to cause cracking without corrosion, the cause was listed as corrosion fatigue.
2. The list shows only the primary failure mechanisms, and many of them actually had multiple physical roots.

THE HUMAN ERROR STUDY

These failure analyses all involved plant site visits and interviews with many plant personnel. In the study, we realized that, because of our various clients' directives, we couldn't guarantee that all the human roots had surfaced but are reasonably confident that a good representation was found.

The study divided the major human errors into six categories as follows:

1. Design — There are two types of design errors, those of omission and those of commission.
   a. An *omission error* is one in which the engineer/designer fails to notice a design flaw that results in a premature failure, such as a stress concentration or an improper material selection.
   b. A *commission error* is one in which the decision is made to use an otherwise acceptable design under conditions that lead to failure. An example of this could be a situation in which the output of a machine that is operating well at 500 units per day is increased to an operating rate of 550 units per day without a serious study and, as a result of the increase, failures occur.
2. Manufacturing — A manufacturing error occurs when a machine that is properly specified and designed is improperly manufactured.
3. Maintenance — Maintenance errors occur when a well-designed and manufactured machine is improperly repaired, or the repaired unit is improperly reinstalled. An example of this is when a millwright decides to slip fit, rather than shrink fit, a coupling that then enables torsional fatigue forces to crack the shaft.
4. Installation — Installation errors occur when a well-designed and manufactured machine is improperly installed.
5. Operating errors — Operating errors occur when a machine is run at conditions outside normally accepted ranges.
6. Situation blindness — The proper word for this is "nescience" but this category could also be called "can't see the woods for the trees." It pertains to those errors that occur when an obvious problem situation is ignored for an extended period of time, culminating in a major failure. (This category deserves special recognition because it is something we all tend to think the "other guy" is responsible for, and we rarely recognize it in ourselves. Yet, there are times when all of us are blind to answers that are obvious to the other 99% of society. We have all seen times when another driver stops at a green light. And we've all seen drivers accidentally drive through red lights. But fortunately we've never done either one of these actions.)

In developing these categories, the first five were relatively straightforward. However, after a review of what had actually happened, we felt that an additional category was warranted. We use the term "situation blindness" to describe those failures in which the group neglect of a well-recognized but relatively minor problem allowed it to grow into an expensive and catastrophic failure. One unfortunate example of this was a 2600-hp reducer failure in a large paper mill. It was in a plant in which procedures said that every operator was responsible for checking the lubricant level on every machine in their area on every shift. The reducer was located on a pedestal that was remote from the operator's control room and awkward to get to, down a flight of stairs, across the operating floor, and up a ladder. It had been leaking heavily for more than 12 months and both the area around the reducer and the sides of the pedestal were saturated with oil. Certainly the direct causes were the leaky seal and the lack of operator attention (failure to replenish the oil), but we felt there also had to be recognition of the general lack of insight on the part of the plant operating and supervisory personnel. When a critical process depends on human reliability, there will be failures.

Another example involves the repeated failure of a 400 kw (300-hp) crusher. When we were called in to do the failure analysis, there were many complaints about the poor design and lack of support from the vendor. Yet, after a couple of days of detailed analysis and long conference calls with the manufacturer and the plant, we found the plant had ignored the poor performance in a previous installation, had a local shop repeatedly repair the machine in a manner the vendor didn't agree with, had installed the machine incorrectly, had tried to use the machine for an application it wasn't designed for, and the vendor had recommended a much larger machine for the application.

The study found a total of 276 major human errors and hundreds of minor contributory causes. The values below indicate the number of times a specific type of major failure cause occurred. Note that there is almost always more than one contributory cause. A typical example of this was a calciner shell failure that was caused by pitting corrosion. The pitting resulted from the combined effects of a welding error during fabrication and an operating error.

The total number in each category shows:

Major Human Errors	Number of Errors
Maintenance errors	93
Design errors	90
Operational errors	34
Manufacturing errors	28
Original installation errors	20
Situation blindness errors	11

Recognizing that the distribution of these major contributors wasn't uniform, we then looked at the actual number of failures affected by each category, i.e., we sorted to see how many involved design errors, maintenance errors, etc. In assessing the major failure causes, the following totals indicate the percentage that the specific groups of errors effected.

Group of Errors	Percentage
Design errors	59
Maintenance errors	38
Operating errors	24
Original installation errors	16
Manufacturing errors	12
Situation blindness errors	9

We began the project with some guesses as to what we would find, but the results of the data analysis were not as expected. To ensure minimal errors, both the analyses and the data summaries were reviewed several times. Looking at the year-to-year summaries, there were variations but they were all within the same general range and we also found others with similar results.

During this same period we were also deeply involved with several true RCAs. These indicate that when a major industrial failure is taken to the point where all the possible causes are examined, there are usually somewhere between 8 and 12 significant roots. The distribution of the physical and human roots tends to fall along the same lines as what we saw in the root cause investigations (RCIs), with the exception that there are usually additional minor roots discovered. The big difference is in the discovery of the multiple latent roots.

LATENT ROOTS

These are the systems or the practices that allow the human and physical roots to exist. For example, most companies make tremendous efforts to ensure good product quality. The standards are very strict, and frequently even one error in a million production units is not acceptable. Yet, many of these same companies don't have receiving inspection programs on their maintenance supplies, and we know from studies that typically 4% of what is received on the maintenance dock is not what was on the purchase order.

Understanding latent roots can be difficult, especially if you are in the midst of them, so a medical example may yield a better viewpoint of how these latent roots effect failures. If you had a heart problem, what type of doctor would you consult? Why wouldn't you go to a dermatologist? The dermatologist has the same premed education as the cardiologist, and then they go through the same 4 years of medical school. But, there is some difference between residency and specialization, and you want to be sure the doctor is correct when it comes to your health. Nevertheless, in many plants, we see a management philosophy where the choice has been to reduce apparent costs, which effectively states that "engineers are engineers" and fails to recognize the differences between chemical engineers, mechanical engineers, mining engineers, etc. Certainly, a chemical engineer can get out a handbook and design a mechanical drive, but what would your reaction be if your doctor practiced out of a handbook? This type of management thinking denies the value of experience and is the consequence of having systems that don't accurately determine the cost of these failures. As a result, they have more design errors and reliability problems in both their production operations and their products.

Some simple examples of management's lack of awareness, each of which cost more than $150,000 (in 2004), concerning engineers are:

1. The chemical engineer who modified the design of a conveyor drive and didn't realize that the position of the drive pinion had an effect on the forces on the conveyor shaft
2. The mining engineer who suffered through repeated fastener failures without understanding how fatigue loading affects bolts
3. The mechanical engineer for a chemical-treating facility who didn't understand intergranular corrosion
4. The design engineer who specified a reducer lubricant that was incompatible with the grease that came in the supplied ball bearings

These latent errors are everywhere and life would be much less challenging without them. However, at times that would be a pleasant change.

HOW THE MULTIPLE ROOTS INTERACT

In looking at failures, there is a general tendency to ignore the latent roots and blame the individual. The result of doing this is that it doesn't solve the problem and doesn't stop it from happening again. There is solid data that shows that the typical plant failure has four to seven physical and human roots. If a failure has six roots and action is taken to correct only one of them, i.e., censuring or penalizing the person involved, what happens to the sources of the other five roots? They come back to haunt and create another failure.

WHY MULTIPLE ROOTS ARE FREQUENTLY MISSED

Early in my career as a reliability engineer, I was told, "One of the real problems with machinery reliability is that it isn't what we know we don't know that hurts us. What is really painful and expensive are those things we think we know that we really don't understand."

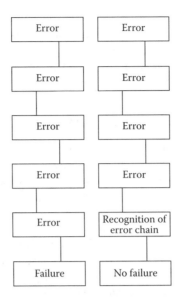

FIGURE 1.1 The error-change phenomenon with the first chain showing the complete failure sequence. In the second sequence, the error chain is recognized, and an appropriate intervention has broken the chain.

Looking at the fact that there are multiple causes to all failures, Charles Latino has often spoken of what he calls the "error-change phenomena." A typical plant failure results as the culmination of a sequence of events. If a person recognizes that this sequence has the possibility of causing a failure and makes a change that breaks the sequence, that failure won't happen. However, when the chain is broken by a random incident, then the probability of the failure recurring is significant. In Figure 1.1, we show the error change phenomenon with the first chain showing the complete failure sequence. In the second sequence, the error chain is recognized, and an appropriate intervention has broken the chain.

In one example, a process pipe in a huge Canadian chemical plant broke, resulting in a spill to an adjoining stream that caused an expensive fish kill. In the analysis of the failure, it came up that they had had two similar failures in the past 10 years. After each of the previous failures, the equipment had been repaired, yet the failure recurred. As in most plants, the normal practice had been for the plant people to look at the failure, repair it as fast as possible, and get back into production. A detailed investigation of the incident using a logic tree found that there were nine major physical and human roots. It was only after the analysis that the plant personnel recognized that some of the causes had been there for years, whereas others were relatively recent. They had a chain of nine events that caused the failure, and any time the nine occurred in the correct sequence, the failure returned. Only by removing several of the roots could the problem be permanently eliminated.

The incident involved the failure of the piping below the storage tank. In the latest event, the pipe cracked between the tank and the discharge control valve, and the flow of ammoniated water couldn't be shut off. Among the major human roots that the analysis found were: the pumps were run until they cavitated and were sometimes allowed to cavitate for long periods of time (on the order of hours); the piping system had a resonance that was excited by the pump vibration; the pipe repair procedures resulted in piping that was inherently much weaker than the original design; the pumps were being improperly repaired; the bases were corroded and greatly weakened; and so on. If any one of these were corrected, the system would be back in operation — and waiting until the next time to fail.

One looks at this failure and says, "How can the plant operators allow severely cavitating pumps to run for hours? Don't they understand the damage that's being done?" Yet, less than 2 months later, we were 1500 mi away, investigating a similar failure in the southeastern USA. Again, the operators allowed the pumps to cavitate for hours and the pumps were destroyed. In both sites, the people were fairly intelligent but had never been taught that severe starvation can destroy a set of pump bearings in less than a day. These are two excellent examples boots of latent roots and the value of education.

THE BENEFITS AND SAVINGS

The cost savings that result from failure analysis can be compared favorably with the benefits of any other effective system analysis. Essentially, we are looking at a problem, understanding in complete detail how and why it occurred, and then trying to prevent it from occurring again. The results should be increased asset availability, increased throughput, and reduced costs. The magnitude of the benefit is proportional to the amount of site support. We have been associated with several failure analysis/reliability programs that were supported only by the maintenance organization and where the carefully documented annual returns were in the range of six to ten times the program cost. However, when supported across the organization, other programs have shown production outputs that have steadily increased over the years with no capital investment and reduced maintenance and operating costs. In an active and well-supported reliability/failure analysis program, it is not surprising to see annual returns of more than fifty times the cost.

The plant where I learned about *The Reliability Approach*™ was over 90 years old with a mature process that had not been changed for years. The annual sales volume of the plant was about half a billion dollars, and it had historically lost around 17% of the production capacity to downtime. During the first six years of this reliability program, despite minimal expenditure on capital equipment and a reduction in both the maintenance staffing and the maintenance budget, we were able to increase effective production time by about 1% per year, and the actual output increased by more than that. (The net benefit was about a 1/3rd reduction in downtime for a process that hadn't materially changed in 40 years.) The following graph shows the actual results from the program (Figure 1.2).

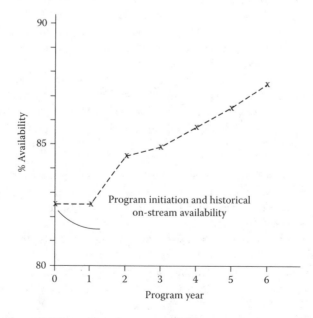

FIGURE 1.2 Equipment availability vs. program year for an effective reliability program.

Industry personnel with extensive experience in improving plant reliability generally think of dividing analyses into three categories in order of complexity and depth of investigation. They are:

1. Component failure analysis (CFA) looks at the piece of the machine that failed, for example, a bearing or a gear, and determines that it resulted from a specific cause such as fatigue, overload, or corrosion and that there were these x, y, and z influences. This type of analysis is solely designed to find the physical causes of the failure.
2. Root cause investigation (RCI) is conducted in greater depth than the CFA and goes substantially beyond the physical root of a problem to find the human errors involved. It stops at the major human causes and doesn't involve management system deficiencies. RCIs are generally confined to a single operating unit.
3. Root cause analyses (RCA) includes everything the RCI covers plus the minor human error causes and, more importantly, the management system problems that allow the human errors and other system weaknesses to exist. It is not uncommon for an RCA to extend to plant sites other than the one involved in the original problem.

Although the cost increases as the analyses become more complex, the benefit is that there is a much more complete recognition of the true origins of the problem. Using a CFA to solve the causes of a component failure answers why that specific part or machine failed and can be used to prevent similar future failures. Progressing to an RCI, we find the cost is five to ten times that of a CFA but the RCI adds a detailed understanding of the human errors contributing to the breakdown and can be used to eliminate groups of similar problems in the future. However, conducting an RCA may cost well into six figures and require several months. These costs may be intimidating to some, but the benefits obtained from correcting the major roots will eliminate huge classes of problems. The return will be many times the expenditure and will start to be realized within a few months of formal program implementation.

One thing that has to be recognized is that, because of the time, manpower, and costs involved, it is essentially impossible to conduct an RCA on every failure. The cost and possible benefits have to be recognized and judgments made to decide on the appropriate type of analysis.

A confusing point about RCA is that some industry personnel consider a CFA as an RCA and some people consider an RCI as an RCA. In our opinion, if a person wants to really know the reasons why a problem occurred, they shouldn't stop looking at some arbitrary point, then say they know all the causes. There will certainly be times when your position or the plant management places a limit on the extent of an investigation. In those cases, the analysis should be accurately described as what it is and not called an RCA.

It is impossible to write something from a distance in a book and say your savings will be $XXX. However, we can say we've never seen a formal failure analysis program where the annual savings have not been significantly greater than the annual program cost. Some of the companies we've worked with have claimed program savings that initially seem outlandish, but consider Figure 1.3. It is a graph of production vs. time for a large manufacturing plant. As you can see there are the daily swings in output while the plant is averaging 2200 units per day. Then, there is a major failure one day, a disaster, when production ceases. Everybody pitches in and works all sorts of hours until production is restored and then the failure analysis begins. This sporadic failure is studied in detail, sometimes with outside specialists, until even the janitors know how much it cost. In the meantime, production has gone back to its daily swings, 1600 units today, 2700 tomorrow. What management fails to recognize is that the annoying small day-to-day failures are costing them over 400 units per day — the difference between their average and realistic capacities. These day-to-day annoyances that cause the rate to swing back and forth are *chronic failures*. They

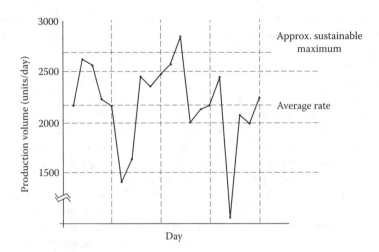

FIGURE 1.3 A plant's production history.

are rarely studied, except by companies truly committed to reliability, and are the source of amazing production increases and cost savings. (Although the preceding numbers are from one specific plant where *The Reliability Approach*™ was implemented, we have worked at several others where the numbers are essentially identical and benefits of 20 to 50 times the program cost have been realized.)

Using Logic Trees

There are many different tools for finding the various roots of failures. These include logic trees, Ishikawa (fishbone) diagrams, and other systems that use an infinite number of combinations as well as a variety of copyrighted formats. Analyzing all of them, we feel that logic trees offer the most efficient method of finding the roots, not only because of their superior flexibility but also because the strict use of the tree requires the verification of each step. This effectively counteracts the most noteworthy weakness of many of the alternatives that allow:

A hypothesis to be stated but not formally confirmed
Opinions to be stated concerning a possible cause, without carefully considering the other
  possible alternatives

Using logic trees correctly requires both the verification of the hypothesis and the consideration of all possible contributing factors. The benefit is a measurable improvement in the accuracy of the analyses and the confidence of the organization.

SUMMARY

In summary, we see that failures have three levels of roots:

1. Physical mechanisms
2. Human errors that result in the physical causes
3. Latent roots that allow the human errors to occur

A number of different types of analyses fall under the umbrella of RCA, and we prefer to look at them as:

- CFA — A CFA that uncovers the physical reasons for failures.
- RCI — An RCI goes deeper than a CFA and finds the major human roots but doesn't find all the human roots and doesn't look at the latent roots.
- RCA — An RCA finds all the contributors to a failure and, although the investigation is costly, typically pays for itself many times over.

Human error is unavoidable, and we all make errors that contribute to failures and accidents.

2 Some General Considerations on Failure Analysis

THE FAILURE MECHANISMS — HOW THEY OCCUR AND THEIR APPEARANCES

As shown in Figure 2.1, there are only four general categories of failure mechanisms — overload, fatigue, wear, and corrosion. All mechanical equipment failures fall into these categories. Though occasionally there may be doubt as to which category a specific mechanism should be ascribed, and sometimes more than one mechanism is involved, all failures can generally be put in the preceding four categories. The following pages will try to take a practical approach and explain the basics of how these mechanisms function and how they can be recognized.

Looking at the difference between the right and left sides of Figure 2.1, it is apparent that the corrosion and wear mechanisms, with very few exceptions, generally remove material from the parent piece until the piece fails from the normally expected stresses. On the other side, the mechanisms on the left half of the chart essentially overstress the material. However, one caution that experience teaches is that in many of the failures there is more than one mechanism at work.

WHEN SHOULD A FAILURE ANALYSIS BE CONDUCTED AND HOW DEEPLY SHOULD IT GO?

It really depends on the possible consequences. When a situation develops in which there is the possibility of a serious injury or serious environmental event, a root cause analysis (RCA) is always warranted. The rest of the time the decision should be based on the costs and the possible plant implications. (Some of the most valuable RCAs in industry have actually been conducted on near misses!) The situation has to be analyzed to determine both the failure cost and how long the part should have lasted, then analyzed to see if the cost of conducting the failure analysis and improving the installation are worth it. Then, depending on the potential, a decision to conduct either a root cause investigation (RCI) or a component failure analysis (CFA) should be made.

Failure analysis is time consuming. To maximize the return on the effort, the analyst should start out with an idea of how long the machinery is designed to last and the possible return on the time invested. For example, in one plant that had 2,200 motor-driven pieces of equipment, the management decided to initiate a program of motor failure analysis. At the start of the program, the average motor lasted just over 5.4 years, and, eight years later, the average motor life was 10.9 years. Looking at the cost benefit from the project (in 2005), assume that the average motor was 25 hp; cost $1,700; and the manpower, record keeping, installation, etc., cost was also $1,700. The annual savings from reduced motor replacements was almost $700,000 every year, and this doesn't consider the production increases, etc., which resulted from the smoother operation. From this, we see that the project had great benefits but a difficult part about any failure analysis is that the benefits should greatly outweigh the costs.

HOW LONG SHOULD IT LAST?

An important key to starting a program is realizing how long many of the common industrial components should last before major maintenance or replacement is needed. Figure 2.2 shows the typical lives, running 24 h/d and 350 d/year, which should be expected with normal design loads.

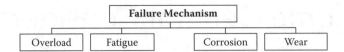

FIGURE 2.1 The four basic failure mechanisms.

Component	Typical expected[1] life - based on current accepted industrial design criteria
Shafts	should never fail
Couplings	
Grid and gear	Should never fail if properly lubricated
Elastomer	Five to ten years
Belt drives	
V-belts	Two to three years
Synchronous	1.25 to 1.5 years
Reducers	Depends on SF (rating service factor).
	Below 1.5 – typically three years
	1.6 to 2.3 – four to eight years
	Above 3 – ten to 15 years
Pumps	
End suction ANSI	Five years
Double suction	Ten years
Fans	
General industrial –	5 to 8 years
mine ventilation –	25 years
Motors (TEFC)	
Below 40 hp	Eight years
40 hp and up	Eighteen years
Chain (power transmission)	Three years (but is highly dependent on lubrication)
Anti-friction bearings	Generally over 20 years

FIGURE 2.2 Expected lifetimes. (Life is defined as "until major maintenance, i.e., overhaul or rebuilding, or replacement is needed!")

Many years ago, a very wise manager said to me "If you get really good at something in maintenance, you're doing it much too often, and it's time for a failure analysis." The undeniable logic behind his statement is that most parts are designed to last many years and should not be changed on a short-term schedule. But, it is commonplace that people put up with frequently changed parts, and we hear of personnel who are proud of their ability to rapidly replace failed components but don't understand why the failures occur.

Before starting a failure analysis, it should be recognized is that this "typical expected life" is not a constant. Economic, operational, and social pressures have had a huge effect. For example, in the 1960s, it was common to specify gear reducers that lasted 12 to 14 years without requiring maintenance, whereas most of the recent applications are frequently less than half of that. On the other hand, typical automobile engine life in the 1960s was around 160,000 km (100,000 mi), and today it is more like 320,000 km (200,000 mi).

Whatever the component is, recognize that the design life is not a hard and fast rule. There are variations in every product that influence its life, and, although the values in Figure 2.2 are typical lives, the distribution of failures almost always follows a bell curve such as that shown in Figure 2.3.

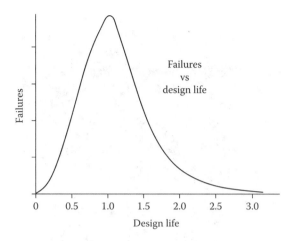

FIGURE 2.3 Typical component life distribution.

One of the difficulties in developing an effective failure analysis program is that it's easy to understand that "it shouldn't happen" when a shaft breaks, but too often failures that don't involve catastrophic damage or an obvious fracture aren't recognized as failures. The areas where there is a lack of awareness of the need for failure analysis commonly include corrosion and wear. These are applications in which occasional replacement or refurbishment is needed but the rework frequency and costs are ignored until a crisis develops with the responsible personnel having lost sight of the fact that their original assignment was to improve plant reliability and reduce costs.

As every reliability professional has seen, another area where the value of failure analysis commonly isn't utilized involves the practice of routinely scheduled outages. Should maintenance be scheduled based on the position of the sun or the moon and end up with monthly or annual outages? What is the real difference between a scheduled outage and a breakdown? The need for maintenance has to be recognized but in many cases the time between these "scheduled failures" can be greatly lengthened by analysis of the need for the various phases of the outage. We have never seen a situation where, after a competent and detailed analysis, the time between outages hasn't been increased by at least 50%.

DIAGNOSING THE FAILURE

It is critical to understand that the failed component always tells an accurate story about what happened. The problem with analysts is that they frequently succumb to human weaknesses and jump to a conclusion before reading the entire story. A good failure analyst needs the observation powers of Sherlock Holmes, an understanding of the operations surrounding the failure, and the ability, diplomacy, and tenacity to question everything.

Looking at the failure face and its surroundings will tell almost everything about the physical causes of the failure, i.e., where it started, the magnitude and direction of the forces involved, how long it took to propagate across the piece, and whether there were any material weaknesses that contributed significantly to it. Sometimes it is more difficult than others to see the clues, and sometimes it is difficult to understand all of them, but they are there. Inspection of the failed agitator shaft shown in Photo 2.1 indicates that it was a fatigue failure, was caused by two-way plane bending, had stress concentrations, and was lightly loaded. Further inspection of the shaft exterior surface will present evidence of the faulty maintenance practices that caused the two-way bending.

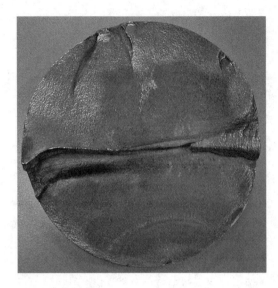

PHOTO 2.1 Fatigue failure of an agitator shaft.

Material weaknesses mentioned earlier is certainly an important topic when dealing with failure analysis, but it is rarely a significant contributor. There is a never-ending tendency to blame the materials, but our failure analyses have found a parent material problem with less than 1% of the industrial failures. However, as soon as people get their hands on a part, whether it is welding, heat treating, or fabrication processes, the number of failures increases.

Most, certainly far more than 80%, of the failures found in the average plant can be solved without resorting to expensive metallography or hiring an outside expert. In these situations, an observant person, or preferably a team of persons, using good lighting and low-power magnification, should be able to see enough of the features on the failed part fracture surface to determine the type of failure and, from that, the forces that caused it. Then, with more investigation, the multiple human errors that culminated in the failure should be apparent.

Failure analyses help the plant find all of the things that contribute to a failure and then do something to prevent a recurrence. But we don't recommend that every failure be investigated and every root always be eliminated. That would be extraordinarily expensive and a waste of precious resources. What we would like to see people do is:

1. Recognize the significant failures and analyze their costs.
2. Identify all the roots, and then intelligently eliminate enough of the major ones to break that error–change linkage and ensure the problem never reappears.

FINDING THE PHYSICAL ROOTS

In conducting the analysis, it is critical that the analyst get to the site rapidly. As time goes on, parts get lost and people's memories either fade or become colored, both of which make the job much more difficult. As soon as possible, visit the site, inspect and collect the pieces, interview the involved personnel, and understand what happened. At the site:

Take lots of photographs of the site and the pieces. Be sure there are enough photos to identify where the pieces started and where they ended up, and try to have a measurement scale in the photos. Also, take photos of the surroundings and the associated parts, then take a few more. Photographs are inexpensive and frequently invaluable.

Discuss the failure with involved personnel and ask them for their opinions about what happened. Try to get to them one or two at a time, and ask them to explain it in their own words. (*Do not* tell them something and ask them if they agree.)

After you've completed these initial steps, step back, look at the failure and think about what could have caused it and what the possible ramifications might be. If there was or is the potential for a large economic loss, human injury, or a serious environmental event, then no effort should be spared in conducting a very detailed analysis. On the other hand, no organization has unlimited assets, and an inexpensive failure deserves an inexpensive analysis.

If the failure is one of those with the possibility of major consequences, the typical steps in the analysis should be:

1. Make a preliminary examination of the failed pieces. What type of failure was it? What basic forces were involved?
2. Collect background data — Look at the installation and the surroundings. What was the normal environment, power usage, operating conditions, etc.? Also, look at the same data for the machines that directly interact with the failed unit. Find out if there have been changes in the last month, year, or other significant intervals.
3. Discuss the failure with involved personnel who work with the equipment but were not there during the initial visit. Again, ask them for their opinions on the causes.
4. Under low-power magnification, look at the pieces, fracture faces, and other surfaces; develop a logic tree that lists every symptom of the failure; and determine the possible causes of the symptoms and the failure scenarios.
5. Conduct more detailed analysis on the critical pieces. This may involve metallography, nondestructive testing, chemical analysis, residual stress analysis, or a variety of other inspection tools used to determine the state of the pieces.
6. From these, determine the physical failure mechanisms and how the failure happened.
7. Then, go further and find the human and latent roots so you can determine how to prevent future occurrences.

Any of these steps can be eliminated or partially eliminated but each step that is skipped increases the chance of making an error. But again, the cost of the analysis has to be balanced against the true cost of the failure.

Of these seven steps, the third step (discussing with the involved personnel) is definitely the most important because we all have different viewpoints. The machine operators will see and hear things that the maintenance personnel don't recognize, the maintenance personnel will see things the engineers don't notice, the operating supervisor will see things the operator doesn't notice, etc. Combining all of them will lead to a more comprehensive understanding of what went on.

When you talk to these people, **listen** to both their words and to what they are trying to say. Many times, they don't have any technical knowledge and may not know the difference between a screwdriver and a hammer — but they have heard and felt things. We have heard pseudotechnical explanations that would make Einstein and Newton turn over in their graves — but they provided great clues to find the causes of failures. Nevertheless, the eyewitness testimony has to be weighed against what the parts say and tempered by the knowledge that many are not accurate in what they report.

PHOTO 2.2 Pitting corrosion resulting from elevated temperatures on an aluminum housing.

In most failure analyses, identification of the failed part is relatively easy. But in complex failures, such as an aircraft crash or the destruction of a large gearbox, finding the piece that failed first is sometimes very difficult, and every piece has to be looked at carefully.

In looking for changes, don't be shortsighted. Sometimes they could have been made 5 or 10 years ago. In one example, we worked with a company that eliminated a dust collection system because they felt the annual maintenance was excessive. After 10 years, the resulting corrosion cost millions to repair. In a second application, the plant practice of leaving out $150 baffle gaskets in a group of heat exchangers resulted in corrosion that required eight tube sheets to be replaced 10 years later — at a cost of $80,000 each. Photo 2.2 shows a pair of aluminum castings. The one to the left failed after the plant put a condensate hose on it to prevent it from freezing the product inside (Why not repair the insulation?), but the condensate increased the casting temperature to over 80°C (180°F), well above the pitting corrosion threshold.

What are the sources of failures? Figure 2.4 is a tree that shows a breakdown of the four major failure mechanisms and their most significant contributors.

Corrosion is probably the biggest source of failures, and total costs owing to corrosion in most of the developed nations have been estimated to be between 3% and 4% of their GNP, a number that has held fairly constant over the past 60 years. There isn't a lot of industrial data but a DuPont survey from the period 1968 to 1970 analyzed their corporate maintenance costs and found the following breakdown (but they said there was only one cause per failure):

1. Corrosion — 55 to 60%
2. Wear — 5 to 7%
3. Mechanical failure
   4. Fatigue — 90% +
   5. Overload — remainder

This list doesn't specifically show a placement for the effects of thermal changes because temperature has a multitude of effects on materials and frequently has a huge effect on causing failures. Thermal stresses contribute to numerous failures yet rarely show any visible symptoms, and we suspect that this lack contributes to the fact that many of the people designing and working with machinery don't understand the forces created by nonuniform heating. Two interesting examples involving temperature differentials and thermal effects are:

1. Measurements on a shiny stainless steel tank roof found the roof temperature was about 55°C (100°F) cooler than the red-oxide-painted external roof beams. The stresses created during weather changes, for example, a hot summer day with a sudden thunderstorm, were close to the yield strength of the metals, and frequent cracking was the result.

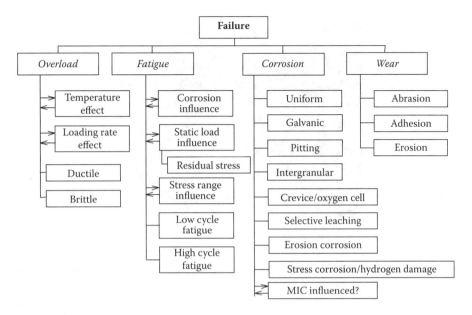

FIGURE 2.4 An expansion of Figure 2.1 showing major failure mechanism contributors.

2. A plant repeatedly pumped 2°C (35°F) cooling water into a large vessel while the vessel wall was still at 95°C (200°F+). Eventually they had serious fatigue cracks.

But these examples just show the thermal stresses. In addition, it has to be recognized that variations in ambient temperatures can have a substantial effect on corrosion rates, can have a major effect on the impact strength of both metals and polymers, and will have an effect on chemical reaction rates. Moreover, large temperature variations can greatly affect both fatigue and wear mechanisms.

INTRODUCTION TO MATERIALS — STRESSES AND STRAINS

The next chapter will cover this in a lot more detail but it is important to understand the type of physical force that caused the failure. If it was a single overload, that would indicate a single unusual event; however, a series of stress cycles shows the machine was being repeatedly worked more heavily than it should have been.

An overload failure tells that some unusual event has happened and the part either bends or breaks almost instantaneously. (We say "almost" because some ductile failures, such as the failure of an inexpensive wire coat hanger, involve the slow collapse of the piece.) On the other hand, fatigue failures result from multiple stress cycles and take place over a comparatively long time. The ability to understand the type of forces (the magnitude and the direction) gives tremendous clues as to the reasons those forces were involved. For example, if a pump shaft failed, the analyst would look at it and find:

If it was an overload failure, such as the piece shown in Photo 2.3, the force was applied in the instant before the piece failed, and some catastrophic event, such as allowing the pump to rotate backwards before starting the motor, should be suspected. Also, the small thumbnail-shaped discoloration at the top of the piece would lead to some suspicion about the material condition.

If it was a fatigue failure, the force that caused the failure was applied over many, many cycles, and a process problem, such as a grossly improper flow rate, would be indicated as a major cause.

PHOTO 2.3 A brittle fracture that originated at a surface defect.

As mentioned earlier, from looking at the fracture face of the rotating agitator shaft shown in Photo 2.1, we can see the failure:

Was a fatigue failure
Took a relatively long time
Was primarily caused by two-plane bending but the loading also had a torsional fatigue
    component
Was very lightly loaded at the time of final fracture, but the load varied over time
Had an appreciable stress concentration on one side

With this information, we can first look at the rest of the assembly and ask how a rotating shaft can be stressed by two-plane bending during operation, then ask how those physical causes came to exist.

Sound interpretation of the surface features will yield a wealth of clues to the failure forces and the modifying conditions and will direct the analyst to the mechanism. From this point, finding the human errors that contributed to the failure can be accomplished with the advantage of a solid foundation.

Determining the Failure Mechanisms

One of the difficulties for a new failure analyst is deciding where and how to begin, i.e., how can they find the physical roots. It becomes easier with practice, but we almost always start by using the chart in Figure 2.5. (We realize that there is no substitute for experience and that this chart will not always lead to the answer. But we also know it works in better than 90% of the plant failures.) This is the first of several similar charts in the book as guides for the beginner — or a refresher for the experienced analyst.

The Plant Failure Analysis Laboratory

What tools are needed in a plant failure analysis laboratory?

Given in the following is a suggested list for a failure analysis lab. The items are listed according to our opinion of the order of importance.

1. *Area* — It doesn't have to be a large room; even the corner of an office will do in the beginning. What you need is a place to conveniently place the failed parts and examine them. Good, glare-free, color-corrected lighting is important. (We have three large fluorescent shop lights with individual controls directly over our main examination bench.)

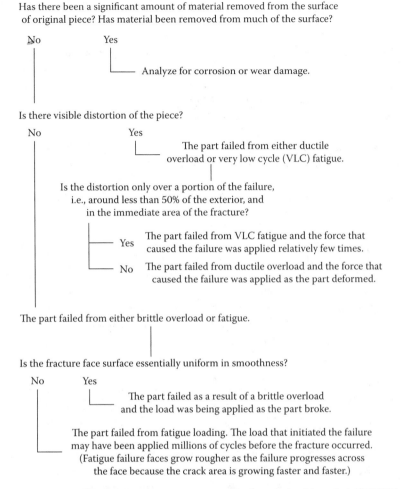

Has there been a significant amount of material removed from the surface
of original piece? Has material been removed from much of the surface?

No　　　　　　　Yes

Analyze for corrosion or wear damage.

Is there visible distortion of the piece?

No　　　　　　　　Yes

The part failed from either ductile
overload or very low cycle (VLC) fatigue.

Is the distortion only over a portion of the failure,
i.e., around less than 50% of the exterior, and
in the immediate area of the fracture?

Yes　　The part failed from VLC fatigue and the force that
caused the failure was applied relatively few times.

No　　The part failed from ductile overload and the force that
caused the failure was applied as the part deformed.

The part failed from either brittle overload or fatigue.

Is the fracture face surface essentially uniform in smoothness?

No　　　Yes

The part failed as a result of a brittle overload
and the load was being applied as the part broke.

The part failed from fatigue loading. The load that initiated the failure
may have been applied millions of cycles before the fracture occurred.
(Fatigue failure faces grow rougher as the failure progresses across
the face because the crack area is growing faster and faster.)

An OVERLOAD failure almost always occurs immediately as the load is applied. FATIGUE
FAILURES happen only after repeated load applications and many cycles.

FIGURE 2.5 Determining the basic failure mechanism.

Depending on the size of the failures in the plant, there may be a need for or access to
material handling equipment.

2. *Digital camera* — A resolution of 2 to 4 megapixels should suffice for most work. It is
important to ensure that the camera has the ability to take close-up photos and that the
area has good glare-free lighting.
3. *Cleaning ability* — After the initial examination and taking of lubricant and contaminant
samples, you should have both a solvent cleaner and an industrial wash sink nearby to
clean the parts for further investigation.
4. *Microscope and light* — A basic binocular microscope with a capability of:
   a. About seven to forty power magnification (Better microscopes have much greater
   depth of field, a tremendous advantage.)
   b. A digital camera direct mounting
   c. A reticule in one eyepiece that would allow for measurement of magnified sections
   The light source should have a variable brightness control and flexible head to enable
   you to shine the beam at various angles across the fracture face.

5. *Lazy Susan* — For ease of rotating bearings and other parts for detailed inspection, a lazy Susan is invaluable. It can easily be made from a turntable bearing with a round plate mounted on it.
6. *Hardness-testing equipment* — There is a direct relationship between hardness and tensile strength, and field testing gives a quick check of a material's properties. (However, it is not foolproof!) There are a variety of scales and testing devices, but we have found the Eqotip electronic scleroscope to be the most valuable because of its versatility. The next hardness tester we would suggest would be a portable Brinell tester.

(In discussing the equipment needs for a plant FA (failure analysis) laboratory, we've assumed the plant has an active predictive maintenance department with tools such as a vibration analyzer, strobe light, infrared scanner, etc.)

When you start the program, we suggest having the metallurgical work conducted by an experienced outside contractor. There are several reasons for this approach:

1. Metallurgy is extraordinarily complex and, unless you have years of education and experience in that field, it is best left to a professional.
2. Metallurgical analysis will be needed on all failures in which there is a chance of serious injury or a significant environmental event.
3. Working with metallurgists and developing a rapport with them will help you to better understand how the material's properties and structure affect failures.

SUMMARY

There are only four basic failure mechanisms — overload, fatigue, corrosion, and erosion. All failures fall into these categories. Deciding how deeply to go into a failure analysis has to be made early on in the analysis and should be based on safety and cost. The part will explain how and why it failed. The difficulty is in interpreting the failure clues.

Overload failures happen instantly as the force is applied. Fatigue cracks require repeated force applications and may take millions and millions of load cycles to progress across the part.

3 Materials and the Sources of Stresses

Some very basic failure analyses can be accomplished without truly understanding the effects of various stresses and the materials they act on. Unfortunately, as failures become more complicated, this lack of knowledge invariably leads to errors. When one starts looking into the more costly or complex failures, the need to understand exactly what is happening becomes incredibly important. Stress causes parts to bend or break but all materials don't react in the same manner. In the real world, a good understanding of stresses and their effects on materials is crucial for accurate failure analysis. (For example, doubling the load causes a beam to deflect twice as much, but doubling the load on a rolling element bearing will cut its life by a factor of eight to ten.) In this chapter, we'll discuss some of those basic definitions and concepts.

STRESS

When a load is put on a part, the part is *stressed*. In Figure 3.1, the load on the block is 4447 N (1000 psi).

In this example, the block is axially loaded, and the piece is in *tension*. In a similar manner, reversing the forces would put the piece in *compression*. To calculate the magnitude of the stress in the block shown in Figure 3.1, the formula

$$s = p/a$$

is used. In this equation, s is the pressure or stress (in MPa or psi), p equals the load (in newtons or pounds-force), and a is the stressed area (in mm^2 or in.^{2}).

If the same block were to be subjected to a bending load or a bending *moment*, the stress would be very different, as seen in Figure 3.2. The block in Figure 3.1 is subjected to pure tension and the stress is uniform across it but, as shown in Figure 3.2, bending puts compression on one side of the piece and tension on the other. To calculate the stress resulting from the bending, the formula

$$s = Mc/I$$

can be used, where M is the bending moment, c is the distance from the centerline to the outer edge, and I is the moment of inertia of the piece.

In addition to tension and bending moments, the piece could also be loaded in a twisting manner with a *torsional stress*, or nonplanar forces that would put a *shear stress* on it. During the design of a part, these loads and the resultant stresses are usually calculated very carefully, and then a significant safety or service factor is added to guard against errors and unexpected failures.

Fractures are always perpendicular to the plane of the maximum stress. At the very beginning of a failure, when only a few grains are involved, the crack may not be perpendicular to the plane of the stress. But once the crack gets big enough to be seen by the naked eye, it always grows in this way, and this fact is invaluable for its ability to point to the source of the load that caused the failure. Examples of this analysis of the fracture plane describing the major load direction are shown in Figure 3.3.

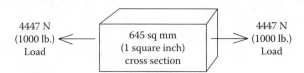

FIGURE 3.1 An example of a tension load that results in a uniform stress of 1000 psi (6.85 MPa).

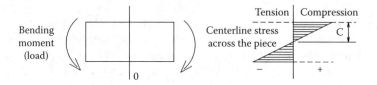

FIGURE 3.2 A bending moment causes varying stress across the piece.

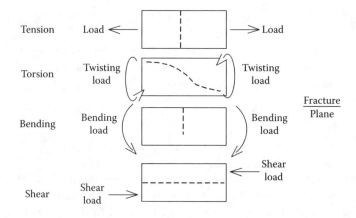

FIGURE 3.3 The four basic loads (forces) and the fracture planes that result from them.

ELASTICITY

All materials are elastic to some degree. What this means is that as a load is put on the part, it stretches as though it were a rubber band. The amount of stretch is directly proportional to the load up to the *yield strength* of the material and the engineering term for this elongation (stretch) is *strain*. Therefore, if a part is stressed, and the load remains in the elastic range for the material, the part will spring back into its original shape when the load is released.

PLASTICITY

But if the part is stressed beyond the yield strength, permanent deformation takes place and the piece becomes plastically deformed. If still more load is placed on the part, it will continue to plastically stretch and eventually fracture when the tensile strength is exceeded.

Figure 3.4 is a stress–strain diagram for two steels. The lower solid line shows mild steel with yield strength of 310 MPa (45,000 psi) and a tensile strength of about 410 MPa (60,000 psi). This means that if the part has a cross-section area of 645 mm² (1 in.²), a load of 309.9 MPa (44,999 lb) can be put on the part before it permanently deforms by 0.2%. If the load on the part is further

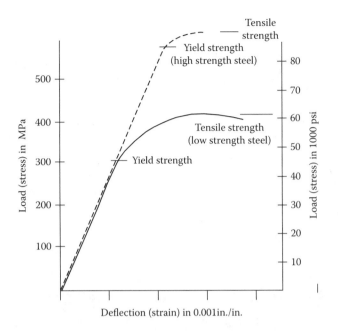

FIGURE 3.4 A basic stress-strain diagram for two different steels.

increased and the yield strength exceeded, plastic deformation takes place. Additional loading causes more deformation, and the part stretches until it breaks at the tensile strength. Next, look at the dashed high-strength steel line. It too is loaded beyond its yield strength and fractures. The difference between the two alloys is that the high-strength steel can carry a much higher load before the yield strength is exceeded. However, the distance between the yield and tensile strengths, i.e., the amount of plastic deformation, is much less in the higher-strength material. From the elongation and plasticity viewpoint, the major difference between ductile materials and brittle materials is that the latter have an extremely short distance between their yield and tensile strengths, and in some materials, such as glass, the distance is essentially zero.

Most engineered designs use ductile materials and use the yield strength as the primary design criteria, applying an appropriate safety factor. The designer knows that a deformed part, a part where the load has exceeded the yield strength, is no longer serviceable and is rarely concerned about the load when the piece breaks in two. (Tensile strength is sometimes used as a design criterion but generally only with brittle materials.)

Photo 3.1 shows a miniature tensile test specimen that has been cut out of a failed piece. This testing is used to determine the yield and tensile strengths and the elongation of the material. In a tensile testing machine, the jaws are locked onto the test piece, and the sample is slowly and steadily loaded in tension. The yield strength is determined by the applied stress when the separation between the two jaws has increased by 0.2%. Measuring the change in distance between two gage points on the bar after the fracture shows the elongation.

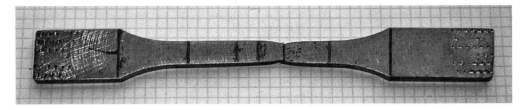

PHOTO 3.1 A tensile test specimen that has necked down (plastically deformed) and fractured.

TABLE 3.1
Elastic Moduli for Different Metals

Material	Metric Units (GPa)	U.S. Units (× 10⁶ psi)
Aluminum	69	10
Beryllium	269	39
Cast iron	90–150	13–22
Copper	105	15.5
Lead	14	2
Monel	175	25
Stainless steel	190	28
Mild steel	210	30
Titanium	105	15
Nylon	7.5–19	1.1–2.8
Polyethylene	48–95	7–14

MODULUS OF ELASTICITY (YOUNG'S MODULUS)

In Figure 3.4, one can see that even though they are different steel alloys, they deflect the same amount when the same load is applied (up to the yield strength). This property, the amount of deflection for a given load, is the *modulus of elasticity*, and is frequently called *Young's modulus*. For all carbon steels, whether the weak inexpensive stuff paper clips are made from or a superstrong heat-treated alloy, the modulus of elasticity is 30×10^6, and they elongate identically. (The reason we sometimes use higher-strength materials is that they will elastically deform much more before taking a permanent set, i.e., before their yield strength is exceeded.)

Other metals have differing moduli as shown in Table 3.1.

Looking at the modulus of elasticity values in the table, one sees that if identical pieces are made out of aluminum and steel and loaded in the same manner, the aluminum part will elongate or deflect three times as much as the steel one. This elongation (or strain) is relatively easy to calculate.

> The modulus of elasticity, E, is expressed in GPa or psi and can be visualized as the load that would cause an elastic part to double in length. Knowing the stress (s), area (a), part length (l), and Young's modulus (E), it is relatively simple to calculate the elongation or strain with the formula:
>
> Elongation = stress × length/area × Young's modulus

TOUGHNESS

In the preceding sections, we discussed the concepts of tensile and yield strengths. In conducting the laboratory tests to find these strengths, the test samples are clamped in a large machine, and the load on the sample piece is very slowly increased. Running the testing machine may take over a minute, starting with no load until the time a part is torn in two. But, in the real world there are times when loads are applied very rapidly, i.e., the parts are impact loaded such as when something is struck with a hammer. It is important to realize that there is little connection between *strength* (with slowly applied loads) and *toughness* (with very rapid load application).

As an example, look at the two bars of the same size shown in Figure 3.5. The bearing-steel bar is almost 10 times as strong as the mild steel bar. What would happen if two rings, 75 mm × 12 mm × 3 mm (3-in. diameter × ½-in. width × ⅛-in. thickness), were made, one from the mild

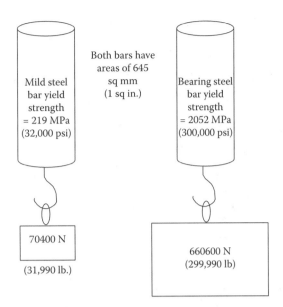

FIGURE 3.5 comparing yield strengths of two common steels.

steel bar and the other from the heat-treated bearing steel, and both were struck hard with a large hammer? The mild steel ring would be badly bent but the hardened steel ring, even though it is much stronger, would shatter like glass. The mild-steel part has the ability to absorb more energy and is tougher than the much stronger bearing steel.

This *toughness* is the ability of a part to absorb energy while fracturing. There are many different toughness testing methods, but the most common in the United States is the *Charpy V-notch test*. Figure 3.6 shows a sketch of a Charpy tester in operation.

The sample is mounted in the vice. Then the hammer is raised to the starting height and allowed to swing freely, hit and break the sample, and continue up to the finish height. If the part were brittle, such as glass, it would absorb almost no energy, and the hammer starting and finish points would be almost identical. On the other hand, some extremely tough nickel alloys are known as "hammer stoppers." For most samples, the hammer would break the bar and swing up to a point a good deal lower than the starting height. This difference in height is carefully measured and multiplied by the mass of the hammer to determine exactly how much energy the sample piece absorbed.

In the temperature range of –75°C to +250°C (–100°F to +500°F), tensile and yield strengths for most steels are relatively constant. But, in many materials the toughness varies tremendously (sometimes by a factor of more than 10) in the range between about –25°C and +100°C (–10°F and 212°F).

> The reason for this change in toughness in some materials is that the fracture mechanism and the energy needed to propagate the crack sometimes change with temperature. For example, with a 0.45% carbon steel, at temperatures over 75°C (150°F), the fracture mechanism is usually *ductile rupture*, where it takes a great deal of energy to tear each metal grain apart. As the temperature decreases, the fracture mechanism gradually changes to cleavage along the grain boundaries and requires much less energy.

FATIGUE

Long ago, in the early days of metal components, people recognized that there were some failures that didn't happen instantly from overload or impact loading. These failures occurred over time in parts that were in constant use and, relating them to tired human muscles, the term *fatigue* was

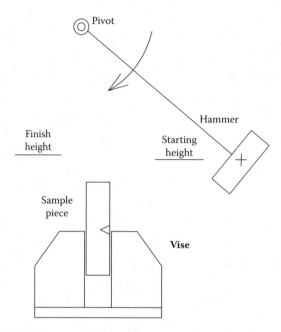

FIGURE 3.6 A Charpy V-notch impact test. (There are similar tests where the piece is not notched.)

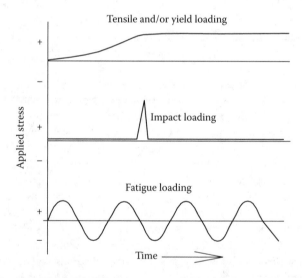

FIGURE 3.7 Comparing the relationship between stress and time for three common loading mechanisms.

used. Figure 3.7 shows a comparison of the three common loading methods, and one sees that fatigue involves alternating tensile forces on a piece. Typically, the long-term *fatigue strength* of a metal is approximately one half of the tensile strength; however, there are many variables that affect the fatigue strength and these have to be recognized.

FATIGUE STRENGTH VS. TIME

The fatigue strength of a metal depends on both how heavily the part is loaded and how many times the load is applied. Figure 3.8 is an SN diagram, where S is the stress and N is the number of load cycles, and it shows how fatigue strength changes with the number of cycles. In the SN

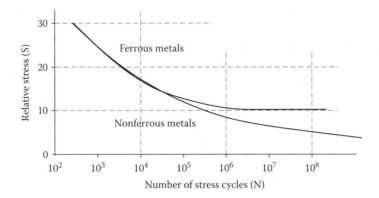

FIGURE 3.8 A typical S–N curve showing how the ferrous and non-ferrus materials react to repeated stress cycles.

diagram, it can be seen that if a part is heavily loaded it can only withstand a relatively few stress cycles, and as the load decreases, the number of stress cycles before failure increases.

This is an essential diagram, and there are several points about it that should be discussed in much greater detail.

1. SN charts usually show fatigue strength as though it was a well-known precise property, when in fact the range between 1% survival and 99% survival is frequently an order of magnitude. This means a part may fail in 10,000 cycles, but another identical part with an identical load might last 100,000 cycles.
2. Generally, the fatigue strength of steel and other metals that have a body-centered cubic (BCC) structure does not decrease after 10^7 stress cycles. On the other hand, the fatigue strength of metals with face-centered cubic (FCC) structures generally decreases continuously. FCC materials do not have a defined fatigue strength. Consequently, handbook fatigue strengths for BCC metals, such as steel, are given for 10^7 or greater cycles, but the fatigue strengths for FCC metals (copper, aluminum, etc.) will always be given with a qualifying number of cycles, for example, 100 MPa (15,000 psi) at 10^6 cycles.
3. Corrosion has a tremendous effect on fatigue strength and continuously reduces it with time. In Table 3.2, one can see that even material seemingly as benign as river water has a serious detrimental effect on fatigue strength. There is no good data that can be used to broadly predict the effect of this deterioration, but some basic guidelines are:
   a. If there is visible corrosion or rusting, the fatigue strength has been reduced.
   b. The longer the corrosion goes on, the more the fatigue strength is reduced.
   c. The stronger the material, the more rapid is the reduction in fatigue strength.
   d. The slower the fatigue stress is applied, the more the strength is reduced.
   e. Aerated solutions, such as sprayed water, cause more rapid reductions than the same material with less available oxygen.
4. Other fatigue-strength-modifying factors that also have to be recognized are:

Size — The larger the stressed volume of a part, the lower the fatigue strength. This factor becomes more important as the component lives grow longer.

Loading type — Axial and torsional fatigue strengths are less than bending fatigue strengths, primarily because a larger volume of the part is stressed.

Surface finish — Surface roughness becomes more important as the strength of the parent material increases, but poorly machined or cut surfaces will reduce the fatigue strength of a part under any circumstances. The rougher or more irregular the surface, the lower the fatigue strength.

TABLE 3.2
Some Corrosion Fatigue Data for Ferrous Alloys

Material	Heat Treatment	Tensile Strength (MPa/ksi)	Type of Loading	Corrodent	Endurance Basis (Cycles)	Fatigue Limit in Air (MPa/ksi)	Corrosion Fatigue Str (MPa/ksi)
Mild steel	Normalized		Rotating bending	River water drip	10^8 cycles	234 MPa/34 ksi	28 MPa/4.1 ksi
0.21% C steel	Annealed	438/63.6	Torsion	Sea water	10^8 cycles at 1300 cpm	197/28.6	26/3.8
0.21% C steel	Annealed	438/63.6	Torsion	Sea water	10^8 cycles at 1500 cpm	124/18.0	34.5/5.0
12.5% Cr steel		895/130	Torsion	Fresh water	2.5×106	225/32.6	110/16.0
18/8 SS		1150/168	Torsion	Fresh water	2.5×106	170/24.6	73/10.6
18.5% Cr	Annealed	690/100	Torsion			215/31.2	170/24.6
0.5 C, Cd plate		990/144	Torsion			178/25.8	45.5/6.6
SAE 1035	Normalized	540/78	Rot bend	6.8% salt solution	107	249/36.2	152/22.0
SAE 1035	Normalized	540/78	Rot bend	6.8% salt solution + H_2S	107	249/36.2	65/9.4

Source: Fatigue Design by C.C. Osgood (John Wiley & Sons, NY 1970).

Surface treatment — Processes that leave the surface with a compressive stress, such as
case hardening or planishing, will increase the fatigue strength of the part. On the other
hand, surface finishes such as hard plating will substantially reduce it.

Temperature — Above about 300°C (550°F), the fatigue strength of steels drops rapidly.

Therefore, when calculating the actual fatigue strength of a material, that is, the fatigue strength
of the part in operation, one has to use:

Actual fatigue strength = Textbook fatigue strength × Size factor ×
Loading factor × Surface finish factor × Surface treatment factor ×
Temperature factor × Corrosion factor

SOME BASIC METALLURGY

Most of the metal used in today's manufacturing plants is relatively low-strength steel. This *mild
steel* is an alloy of iron and carbon. If it were just iron, it would be extremely weak and carbon,
up to about 0.3%, is added to strengthen the alloy, but there is little else in the way of alloying
elements.

Figure 3.9 shows the basic structure of an iron crystal. Each of the "Fe" symbols represents
an iron atom and, because it is a metal, the crystal structure is regular and uniform. When carbon
atoms are added to the structure, they distort it to a degree such that the crystal becomes much
stronger. Unfortunately, there are times when additional atoms are added to the structure and the
result is a general weakening. Looking at Figure 3.10, it is relatively easy to see that a vertical
stress across the plane of the impurity would readily lead to a failure.

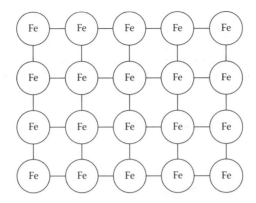

FIGURE 3.9 An idealized representation showing iron atoms in a perfect crystal.

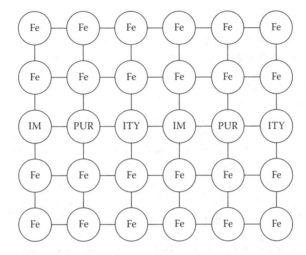

FIGURE 3.10 Adding an impurity weakens the crystal structure.

With the knowledge that the basic structure of steel (and almost all other metals) consists of a metal crystal structure strengthened by alloying elements, one can then begin to look at the major variations and their results.

CARBON STEELS

Plain carbon steel is an alloy of carbon and iron. The largest portion of this mix is iron, but on its own, without the addition of small amounts of carbon, iron is extremely weak. Carbon acts as a strengthening agent, and the amount of carbon can vary from almost nothing to up to 1.1%. The real value of carbon is that it gives steel a huge improvement in mechanical properties compared to those of basic iron. Very-low-carbon steels, such as AISI 1010, are relatively weak, while higher-carbon steels such as AISI 1045 or AISI 1095 are much harder and stronger.

A general division of these carbon steels is:

Low carbon — Up to 0.29% carbon
Medium carbon — From 0.3% to 0.6% carbon
High carbon — From 0.6% to 1.1% carbon

(Between 1.1% and 1.7% carbon is a material usually called semisteel, which shares many of the properties of cast iron and is used in the cast form.)

In addition to carbon, some of the common elements in low-carbon steel are silicon, manganese, phosphorus, and sulfur. The silicon (up to 0.6%) and manganese (up to 1.65%) are added to help remove oxygen from the molten mix. (If the oxygen isn't removed, it combines with other materials and forms oxides, particles that weaken the finished product.) The phosphorus and sulfur are contaminants, and their levels are closely controlled.

The advantages of plain carbon steel are that it:

- Is relatively low cost
- Does not take very sophisticated equipment to make
- Can readily be worked (or formed)
- Offers a great improvement in strength and ductility over wrought iron, cast iron, and most of the nonferrous alloys

The problems are that it:

- Is not as strong as we would like it to be
- Weakens rapidly at higher temperatures
- Cannot be heat treated to be hard and strong without becoming very brittle
- Doesn't have very good corrosion resistance

Low-alloy steels are designed to overcome the first three of those objections and have additions of elements such as nickel, molybdenum, boron, vanadium, and chrome, up to a total of about 6%, so the properties of the plain carbon steel can be improved. Generally, adding these elements enables larger pieces of steel to be hardened more successfully while adding toughness. (There are also special additions that can be made to improve qualities such as low-temperature toughness and high-temperature hardness, but they are beyond the scope of this text.) In the following, we'll define the terms used in Table 3.3 and discuss some of the basic types of steel.

In a pretty simple comparison, making a metal alloy such as steel is similar to making bread; you assemble the ingredients, mix them in a certain way, then cook them for a given time. There are the essentials, such as flour and liquids (iron and carbon), then a number of additives that give the mix a special flavor, like raisins, maple syrup, chocolate, etc. (nickel, chrome, molybdenum, etc.). The way the bread is kneaded changes the end result, as does the way it is heated. With steel, by mixing the correct additives at the proper time and using either heat or cold working techniques

TABLE 3.3
Showing Several Common Steel Alloys and the Major Chemical Constituents

AISI#	Carbon	Manganese	Chrome	Nickel	Molybdenum
1018	0.15–0.20	0.60–0.90			
1020	0.10–0.23	0.30–0.60			
1035	0.32–0.38	0.60–0.90			
1045	0.43–0.50	0.60–0.90			
4140	0.38–0.43	0.75–1.10	0.80–1.10		0.15–0.25
4340	0.38–0.43	0.60–0.80	0.70–0.90	1.65–2.00	0.20–0.30
8620	0.18–0.23	0.70–0.90	0.40–0.60	0.40–0.70	0.15–0.25

to strengthen the alloy, a tremendous range of strength and ductility properties can be controlled to meet our needs.

In analyzing the choices as to what alloy should be used, the component strength isn't the only thing to consider. In addition, size, final cost, ease of fabrication, reliability, etc., all have to be looked at. If all this sounds somewhat confusing to you, you're not alone and shouldn't feel bad. In one of the books in our library, there is a listing of sun gears from a series of automotive transmissions. One would think there wouldn't be a big variety in the materials chosen because of the similarity of the applications. However, there were: 7 manufacturers, 11 transmissions, 9 different steels, 2 heat treatments, and 10 different hardness specifications. Three of the manufacturers had more than one transmission and in *no* case were the specifications the same. Metallurgy and materials science are very complicated, and if you feel a little confused it seems like you have good company.

Before we go further, one of the truly important concepts to understand is that of *hardenability*. There is a direct relationship between hardness and tensile strength, and the harder a part is, the stronger it is. This property of hardenability describes how readily steel can be hardened and, after cost, really drives much of material selection. It's important to realize that hardenability involves both the depth and distribution of the hardness that we get when a part is heat treated, and many of the alloying elements mentioned in the following are used to improve hardenability. (But in all this discussion, don't forget the difference between strength and toughness. Hard materials are frequently very brittle, and most have very poor toughness. An example of this can be seen by looking at high-strength ceramics that are typically very hard but don't have good toughness properties.)

Iron and its Alloying Elements

The principal alloying elements for iron and some common applications are:

Carbon converts pure iron to steel. Up to a little over 1%, the higher the carbon percentage, the harder and stronger the steel. Above about 1.7% carbon, the alloy is usually called *cast iron*.
Chromium increases the ability to heat-treat steel and increases the depth of hardness. From:
    0.3 to 1.6% — It forms a low-alloy steel
    10 to 27% — It forms stainless steel. (There are some inexpensive grades of stainless steel with less than 10% chrome and marginal corrosion resistance.)
Manganese, which deoxidizes the mix and also improves response to heat treatment
    0.3 to 1.9% — Low-alloy steels
    10 to 15% — Work-hardening steels (Hadfield steel)
Molybdenum increases toughness and depth of heat treatment
    0.08 to 0.6% in carbon and low-alloy steels
    0.6 to 4.0% in stainless steels, where it is also used to increase pitting resistance (up to 6% in some specialty stainless steels).
Nickel increases strength, toughness, and impact resistance
    0.4 to 3.75% in low-alloy steels
    3.5 to 36% in stainless steels
Silicon deoxidizes and improves tensile strength.
    Normally in the range of 0.2 to 0.35% in low-alloy and carbon steels, but it can range up to 1%.
    From 1.8 to 2.2% in specialty silicon manganese and electrical steels.

Many of these elements improve the response to heat treatment, and this means that the hardened portion of the steel extends to a greater depth. For instance, if a 3-in.-diameter plain carbon (0.4% C) steel bar is heated in a furnace and then quenched, the result is a hardened layer only 3 mm (1/8 in.) deep. But if the bar is of low-alloy steel, such as AISI 4340, with chrome and nickel in it, the hardened area will extend all the way to the center of the bar, and the resultant structure will be both stronger and tougher.

Other alloying elements sometimes seen include aluminum, boron, calcium, columbium, copper, lead, sulfur, tellurium, titanium, tungsten, and titanium. Lead, calcium, and sulfur don't really alloy with iron, but form long "stringers" that help the steel to machine more easily. The others are used to improve specific qualities in the steel.

AISI NUMBERING CODE

The AISI code is commonly used to identify carbon and low-alloy steels. There is another system called the *Unified Numbering System* (UNS) code, which can also be used, but today it is most commonly used to identify stainless and specialty steels. The Society of Automotive Engineers (SAE) system, another cataloging method, is very similar to the AISI numbering but has slowly fallen out of general use in the last decade or so.

The AISI system uses a four- or five-digit number to identify the steel, and understanding the sequence will help tell what the steel constituents are. For example:

AISI 1045 — The first two digits "10" show it is plain carbon steel, and the last two "45" state that it contains 0.45% carbon. (Including tolerances, the allowable carbon range is actually 0.43 to 0.50%.)

AISI 4130 — The first two digits "41" indicate there are chrome and molybdenum additions. The "30" shows the steel has been alloyed with 0.30% carbon.

The most common ranges of these AISI steels are:

10xx — Plain carbon steel, nonresulphurized

11xx — Plain carbon steel, resulphurized

12xx — Plain carbon steel, resulphurized and rephosphorized

41xx — Low-alloy steel, chromium (0.50, 0.80, or 0.95%), and molybdenum (0.12, 0.20, or 0.30%) additions

43xx — Low-alloy steel, nickel (1.83%), chromium (0.50%), and molybdenum (0.25%) additions

46xx — Low-alloy steel, nickel (0.85, 1.83%), and molybdenum (0.20, 0.25%) added

5xxxx — Specialty bearing steels, carbon (1.04%), chromium (1.03 or 1.45%)

61xx — Low-alloy vanadium steel, chromium (0.60 or .95%), vanadium (0.13% minimum)

86xx — Low-alloy steel, nickel (0.55%), chromium (0.50%), and molybdenum (0.20%) additions

xxBxx — Indicates a specialty steel in which boron is added to improve the hardenability of the steel and the action of the other alloying elements.

There are several other popular numbering systems, for example, most of the major industrial countries such as Great Britain, Japan, and Germany have extensive systems for identifying materials. In the United States, the ASTM has a series of standards that cover everything from boiler tube materials to tennis shoes and are generally application specific, such as steel for pressure vessels and bolting for high-temperature applications. Many materials are formulated to meet several different organizations' standards, so plain carbon steel might meet an AISI specification and also meet ASTM specs as well as DIN and JIS standards.

PLAIN CARBON STEEL — THE BASICS

1010, 1018, 1020 — Low-carbon steels of low strength, used for general structural and mechanical parts in which "steel" is needed, also frequently used for case hardening

1035, 1042, 1044, 1045 — Medium-carbon, direct hardening steels used where stronger, more wear-resistant materials are needed

1095 — High-carbon direct hardening steel, frequently used for drills and hardened tools

Free-machining steels — These steels have tensile and yield properties similar to their plain carbon steel counterparts, but are made with additives that make machining easier and less expensive. From a reliability viewpoint, one problem with these steels is that they usually don't have as good fatigue properties as the plain carbon varieties, because the same features that make them free machining and allow the chips to break up easily also cause small stress concentrations. (The fatigue strength is frequently decreased by 5 to 8%.)

1117 — Free machining, resulphurized steel, frequently used for case hardening

1137, 1141, 1144 — Medium-carbon, direct hardening, resulphurized free machining steels

1214 — Low-carbon screw machine stock, resulphurized and rephosphorized

Low-alloy steels — Steels that have the ability to be made stronger and tougher because they have a good response to heat treating.

4140, 4142, 4147, 4150 — Direct hardening low-alloy steels with chrome and molybdenum

4340 — Nickel–chrome–molybdenum direct hardening steel used where exceptional strength and toughness is needed

4617, 4620 — Nickel–molybdenum case/surface hardening steel

6120, 6150 — Chrome–vanadium steels

8617, 8620 — Nickel–chrome–molybdenum steels used for surface hardening

52100 — High-carbon bearing quality steel with chrome additive

When a piece of equipment is designed, the engineer doing the work typically looks up a table, such as Table 3.4, to find a material that has the needed mechanical properties. These tables can be found in numerous texts and also in manufacturers' literature. However, it must be understood that poor heat treatment can create an alloy that is weaker than the strength listed for the hot rolled metal. Also, one must consider whether the value listed the typical or the minimum for that situation. If you have any questions, don't hesitate to call a local metallurgist or the vendor's metallurgist for advice when choosing an alloy.

UNDERSTANDING STEEL TERMINOLOGY AND MATERIAL DESIGNATIONS

Continuing on the cake mix analogy and all the variations that can be made for special occasions. Well, the same thing applies for steels, and there are a multitude of alloys available — just as there are many different bread and cake mixes available.

Cast materials are poured from a furnace and formed into the finished shape while they are molten. A cast pipe flange involves pouring molten metal into a mold and allowing it to cool. The piece comes out of the mold looking much like the finished product, except for some final machining.

Wrought materials are made from a material that is molten in the furnace, poured into a
shape, then physically deformed into the final shape. For a wrought pipe flange, the molten
material is poured from the furnace into a large rectangular billet that is then rolled (hot
forged) into a thinner plate in a number of steps. When the desired thickness is reached,
a smaller piece is cut from that thinner plate, and the flange is machined out of it.
Welds are cast materials that are frequently used to join two wrought materials.

By the nature of the process used to produce them, most cast products have properties that are
relatively uniform in all directions, i.e., they are not *isotropic*. On the other hand, with wrought
and forged products, because the processing deforms them and results in grains that are stretched
in one plane more than the other two planes, they have more strength in the rolling direction than
across the plane of the rolling. Also, because of uniformity and other metallurgical phenomena,
there is frequently some advantage to cast products where corrosion is involved, and there is
generally an advantage to forged products when impact loading is involved.

Hot rolled materials are formed while the basic material is at greatly elevated temperatures and
the resultant product does not exhibit much grain elongation or deformation. *Cold drawn* (also
called *cold worked* or *cold rolled*) materials are rolled out while the pieces are at relatively low
temperatures, the resultant grain structure is elongated in the rolling direction, and the pieces are
harder and stronger than if they had been hot rolled. (These cold drawn materials are *strain
hardened*.) *Q & T* stands for "quenched and tempered" and tells you that the material has been
heat treated and is stronger than an annealed or hot rolled product. *T, G, & P* stands for "turned,
ground, and polished" and usually refers to high-quality precision-machined shafting material.

Cast Iron

Cast iron is a mixture of iron and carbon with carbon percentage above 1.7%. It comes in many
strength levels and metallurgical structures.

The manufacture of cast iron is an ancient practice and there are many "standards," including
gray cast iron, ductile iron, white cast iron, malleable iron, nodular iron, etc. Each of these general
groups has different properties resulting from the foundry and heat-treating practices. We've seen
huge castings made around the turn of the last century that were as weak as 8,100 psi and more
modern ones as strong as 100,000 psi.

Stainless Steels

There are five different families of stainless steels, and their basic divisions and properties are as
follows:

Austenitic — These are the familiar 300-series stainless steels. They typically contain large
amounts of chromium and nickel for corrosion resistance and some have other additions
such as molybdenum for pitting resistance. They are not magnetic and cannot be thermally
heat treated but respond well to cold working and are readily weldable. Types 302 and
304 are the most common.

Ferritic — These are some of the 400-series stainless steels and have high chromium contents
but little else in the way of alloying elements. They are magnetic, cannot be hardened by
heat treatment, and can only be moderately strengthened by cold working. They have good
ductility and excellent resistance to stress corrosion cracking. The most common ferritic
stainless is Type 430.

Martensitic — These are thermally hardenable, magnetic, straight chromium alloys with
moderate corrosion resistance. Their ductility (in the annealed state) and corrosion resistance

is comparable to the ferritic alloys but the ability to be thermally hardened to tensile strengths greater than 1,400 MPa (200,000 psi) makes them valuable for applications in which good corrosion resistance and strength are required. Common alloys include Type 410, which is used for general-purpose applications and Type 440C, which is primarily used for cutlery and industrial knives.

Precipitation hardening (PH) — These magnetic alloys typically contain chromium and nickel as their principal alloying elements, but they also have small quantities of columbium, aluminum, copper, and/or tantalum as their precipitating agents. Their primary advantage is that parts can be machined in the annealed state and then precipitation hardened at relatively low temperatures. Their corrosion resistance and ductility are comparable to the similarly hardened martensitic alloys. Common grades include "13–8" (13% chrome and 8% nickel), "15–5," and "17–4."

Duplex — The duplex stainless steels have a structure that is a combination of austenite and ferrite, a combination of the good corrosion resistance of the 300-series austenitic and the stress corrosion cracking resistance of the 400-series ferritic stainless steels. The result is an alloy that has both good corrosion resistance and high strength. They cannot be thermally hardened and can be welded, but temperature control is very important. The most common grade is Alloy 2205, although there are several others plus specialty heat exchanger tube materials.

STRENGTHENING METALS

There are two ways of strengthening metals — heat treating and cold working. Some metals, such as the common steels, can be strengthened by either method, but others, such as austenitic stainless steels, can only be strengthened by cold working. In the following, we discuss these alternatives and how and why each is used.

HEAT TREATING

When steels are hardened by heat treating, the basic process is that the steel is heated to a given temperature, then quenched by cooling it rapidly in air, oil, or water. (The specific quenchant, air, oil, or water, depends on the type and size of the steel and the desired properties.)

To get the preferred microstructure, the steel should have at least 0.3% carbon. Low-carbon steels, such as AISI 1010, 1020, and 8620 don't have this much carbon. They can be hardened to some degree, but to get the optimum metallurgical properties carbon has to be added. The way this is usually done is that they are heated in an oven surrounded by a high carbon atmosphere. The steel gradually absorbs the carbon, then the parts are removed from the furnace and quenched. The result is a hard exterior shell over a much softer core. There are several terms for the process and the more common ones are *case hardening, surface hardening,* and *carburizing*. Sometimes when small quantities of parts are hardened, instead of putting them in an atmospheric-controlled oven, the heat treater will coat the piece with a compound that releases carbon as it is heated. There are other surface-hardening procedures such as nitriding and carbonitriding are also used.

Higher-carbon steels, those with more than about 0.3% carbon, can be directly hardened, i.e., they don't have to absorb additional carbon and don't have to be put in a high-carbon atmosphere to be transformed to a much harder material. These directly hardenable alloys are heated and then quenched to obtain the hardness. Most often this is done with alloys in which the resulting hardness is essentially the same throughout the part. These are called *through hardened* parts and don't have the sharp delineation between hardened and unhardened areas seen in surface-hardened pieces.

The choice of hardening methods, i.e., surface hardening vs. through hardening, is usually based on production costs but the product usage is also considered. Case hardening is typically used on things such as automotive gears or wear components in which high strength or wear resistance is needed on a portion of the piece. Through hardening is usually chosen for pieces, such as machine shafts, where increased strength is needed throughout the component. (In general, where an entire component is to be hardened, through hardening is less expensive than case hardening.)

Some of the terms involved with heat treating are:

Annealed — Heating the part to a specified temperature, around 1700°F for steel, then controlled cooling to reduce the hardness and improve machining.

Normalized — Heating to a temperature above the transformation range, then cooling in air to a temperature well below the transformation range.

Transformation temperature — The temperature, for most steels, is around 1350°F, above which the grain structure of steel changes from a body-centered cubic to a face-centered cubic.

Quenched — After a part is heated up to a given temperature, it is rapidly cooled. This is called *quenching*, and different quenchants are used such as oil and water to obtain different cooling rates and different material properties.

Stress relieved — Heating steel to a temperature to allow for internal reorientation of the structure to reduce stresses that usually result from processing. Most steels are heated to 620 to 650°C (1150 to 1200°F) for 1 h/in. of thickness, then slowly cooled. (Stainless steels have to be stress relieved at approximately 1000°C [1800°F] to avoid sensitivity to corrosion problems.)

Carburized, carbonitriding, nitriding, and cyaniding — All refer to surface-hardening methods that are usually done in an oven and result in a thin hardened layer over a softer core. Carburizing involves carbon. Nitriding and cyaniding both use nitrogen compounds. Carburizing usually gives a thicker case but is more expensive.

Through hardening — Refers to a part that is essentially the same hardness throughout (as compared to surface or case hardened pieces.)

Flame and induction hardening — Are surface-hardening methods that don't require an oven. Flame hardening is frequently used in low-quantity processes but can also be used in higher-quantity production machinery. Induction hardening is usually a relatively high-quantity production process.

Tempering — Is the procedure in which a part is reheated to a lower temperature immediately after hardening. The primary reason for tempering is to reduce internal stresses and add toughness to the material. In case of hardened products, tempering results in a slight decrease in hardness with a substantial reduction of the stresses across the interface between the case and the core.

COLD WORKING

As mentioned earlier, a second way of strengthening some metals is by cold working. This involves physically deforming the grains at temperatures well below the transformation temperature and results in an elongated, strain-hardened grain structure.

Austenitic stainless steels are particularly good examples because they can't be thermally hardened, but cold working will result in tremendous increases in the yield and tensile strengths. For example, 304 stainless has a yield strength of about 210 MPa (30,000 psi) in its annealed condition, and when fully cold worked, it can reach as high as 1,200 MPa (170,000 psi).

Cold working, cold drawing, work hardening, and *strain hardening* mean the same thing and refer to the same processes.

TABLE 3.4
Reviewing the Typical Physical Properties of Some of the above Alloys with Different Processing Methods

AISI #	Condition	Tensile Strength (MPa/ksi)	Yield Strength (MPa/ksi)	Elongation (%)	Hardness (BHN)
1018	Hot rolled	380/55	220/32	25	115
1020	Hot rolled	380/55	205/30	25	115
1035	Hot rolled	495/72	275/40	18	145
1045	Hot rolled	565/82	310/45	16	160
1045	Cold drawn	625/91	530/77	12	180
1045	Q & T	825/120	620/90	18	240
4340	Q & T	1250/180	1100/160	15	360
8620	Hot rolled	615/89	450/65	25	188
8620	Cold drawn	705/102	585/85	22	210

Source: Data from several industry sources.

Most of the materials that respond well to work hardening cannot be heat treated to increase their tensile strengths. Two excellent examples of this are the 300-series stainless steels (i.e., austenitic stainless steels) and copper alloys.

Carbon steels can be cold worked, as can be seen by the "Cold Drawn" data on Table 3.4. The increase in tensile strengths can range from a minimum of about 5% up to a practical maximum of about 50%.

THERMAL EXPANSION

Any material (solid, liquid, or gas) will expand on heating. This is because there is increased molecular activity. Some materials, such as gases, expand exactly in proportion to the temperature change (as defined in degrees Kelvin). However, with steels and other rigid materials, there are *coefficients of expansion* that can be used to determine how much a material will change in any dimension at a given temperature. For instance, around room temperature, low-carbon steels grow (or shrink) by 0.0000011 cm for each cm of length for each degree Celsius (0.0000063 in. for every inch of length for every degree Fahrenheit) temperature change. So, if there is a 50 cm (19.7 in.) long bar at room temperature and it's heated to 225°C (437°F), the change in length can be found by:

$$\Delta L = \varepsilon L \Delta t$$

where the change in length is ΔL, the length is L, the change in temperature is Δt, and the coefficient of expansion is ε. Therefore, if it starts out at an ambient temperature of 20°C (68°F), the change in length is:

metric $\Delta L = 0.0000011$ cm/cm/°C $\times$ 50 cm $\times$ 205°C = 0.011 cm

US $\Delta L = 0.0000063$ in./in./°F $\times$ 19.7 in. $\times$ 369°F = 0.046 in.

Another example that gives a substantial insight into the deformation and residual stresses caused by welding is to calculate the amount of contraction that occurs in a steel weldment between the point where the weld puddle solidifies and room temperature. An alloy such as steel really doesn't freeze at a single temperature like water, but let's assign the "freezing point" as 1425°C (2600°F), then calculate the shrinkage in a 61-cm-long (24 in.) weld as the part cools to room temperature. With the length as 61 cm (24 in.), the change in temperature as 1405°C (2530°F), and the coefficient of expansion 0.000011 cm/cm/°C (0.0000063 in./in./°F),

$$\Delta L = \varepsilon L \Delta t =$$

metric $\Delta L = 0.00000011$ cm/cm/°C $\times$ 61 cm $\times$ 1405°C $= 0.94$ cm

U.S. $\Delta L = 0.0000063$ in./in./°F $\times$ 24 in. $\times$ 2530°F $= 0.38$ in.

The coefficients of expansion of many metals change somewhat with temperature.
Table 3.5 shows various expansion coefficients, and we can see from it whether a part is made from:

Aluminum — The expansion would be twice that of carbon steel.
300-series stainless steel — The expansion would be about 1.5 times as much as steel.
Plastic — The expansion could be ten times as much as steel.

Wind chill factor does not apply to materials. (It only applies where there is evaporating moisture, such as human skin or water-wetted components.)

TABLE 3.5
Coefficients of Expansion for a Variety of Materials

Material (at Room Temperature)	Coefficients of Expansion	
	in./in./°F	cm/cm/°C
Steel	0.0000063	0.0000113
Aluminum	0.000013	0.000023
Brasses (range)	0.0000096/0.000012	0.000017/0.000023
Brick	0.0000031	0.0000056
Cast iron	0.0000056	0.000010
Copper	0.0000098	0.000018
Glass (average)	0.0000044	0.0000079
Lead	0.000016	0.000028
Nickel	0.0000086	0.000015
Plastics (range)	0.00002/0.00009	0.000036/0.00016
Stainless steels (typical)		
300 series	0.0000093	0.000017
400 series	0.0000070	0.000013
Stone (range)	0.0000031/0.0000079	0.0000056/0.0000014
Titanium	0.0000053	0.0000095
Zinc	0.000015	0.000027

Notes: These coefficients apply in the temperature range between –20°C and 150°C (0°F and 300°F). For other temperature ranges, there are other coefficients. For instance, when steel is heated from 700°C to 750°C (1300°F to 1400°F), it actually shrinks!

There are many generic or slang terms used for metals in industry. Some of the more common ones are:

Black iron — As in "black-iron pipe" is ordinary low-carbon mild steel.

White metal — Was originally a term used to describe any of several tin, lead, and antimony alloys used for Babbitt bearings; now frequently used to refer to any cast aluminum, zinc, and magnesium alloys.

A 36 plate — Refers to ASTM A 36 steel plate. A low-carbon, good-quality, mild steel with minimum yield strength of 36,000 psi (250 MPa).

Wrought iron — Many years ago it was a very-low-carbon steel that was not very strong but easily worked. Today, it is a generic term for mild steel.

Cast steels — Can have basically the same properties as wrought steels with the same chemistry.

Low-alloy steels — Steels with less than 6% alloying elements.

High-alloy steels — Stainless and other specialty steels generally with more than 10% alloying elements.

SUMMARY

A basic understanding of metallurgy is critical to effective failure analysis.

All steels deflect the same amount with a given load. The advantage of going to stronger materials is that they deflect more before permanently deforming. The downside of these stronger materials is that there is a narrower window between the yield strength and tensile strength.

When a part fails, the fracture plane is always perpendicular to the plane of maximum stress.

Looking at the different loading mechanisms, impact loads, steady-state loads, and fatigue loads, all rely on different material properties.

The major benefit of alloying elements is that they improve the toughness, hardenability and corrosion resistance of the parent metals.

4 Overload Failures

There are different basic systems for classifying failure and, as outlined in Chapter 2, we've divided them into four categories — overload, fatigue, wear, and corrosion. We'll start by reviewing the major features of each of these categories, and then investigate them in detail as the failure diagnoses are applied to specific parts.

INTRODUCTION

Overload failures happen almost instantaneously. When a glass bottle falls to the floor and shatters, the glass is overloaded, a brittle fracture occurs, and the cracks grow at an incredible speed. However, with ductile overload failures, similar to the collapse of an inexpensive wire hanger, the metal bends, buckles, and fails, not as rapidly as a brittle fracture but much more rapidly than fatigue, wear, or corrosion take place.

Overload failures can be divided into two classifications, ductile failures and brittle failures, because of the difference in their appearances. For example, identical overload forces can act on two pieces, one a ductile material and the other a brittle material, and result in very different fracture appearances. The ductile part bends and distorts, and there is typically a great deal of elongation between the yield point (when the first visible distortion happens) and the final fracture. On the other hand, with a brittle fracture, the yield and tensile strengths are almost identical, so there is no warning of a failure, no measurable elongation, and the part suddenly breaks in two when the load becomes great enough.

Glass is an example of a brittle material and marshmallows are an example of a ductile one, but not in many materials is this property obvious. Some of the common materials used in industry can be divided as follows:

- Ductile materials — Mild steel (unhardened steel), most aluminum alloys, most copper alloys, titanium, austenitic stainless steel, many plastics, etc.
- Brittle materials — Ceramics, cast iron, concrete, hardened steels, glass, some aluminum alloys, wood, some plastics, etc.

When looking at overload failures, it is important to recognize the differences in appearance between the brittle and ductile failure modes, and Figure 4.1 shows how, with the same loads, the fracture appearances of ductile and brittle fractures are very different.

An example of the difference in appearance that can be seen in two identically loaded parts is shown in Photo 4.1. These are three elevator chain rollers used in industrial bucket elevators, and they function very much like roller chain rollers. The center piece is as they were supplied and the pieces on either side were from different batches of rollers. The one on the left was hardened and has fractured longitudinally, whereas the one to the right has tremendous ductile deformation. Identical loads have thus caused very different fracture appearances.

Photo 4.2 is an excellent example of the difference between ductile and brittle torsional overload failures. It shows two torsional overload failures. The piece on the upper right is from a utility truck axle, and this brittle fracture started at a large metallurgical defect and occurred only a few thousand miles after the truck was delivered. Below and to the right are two pieces from the ductile overload failure of an inexpensive ½-in. socket wrench extension.

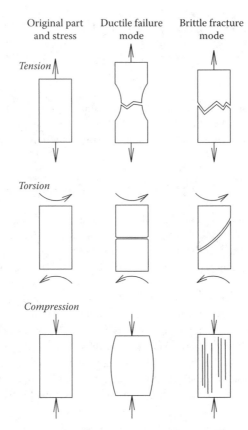

Original part   Ductile failure   Brittle fracture
and stress   mode   mode

Tension

Torsion

Compression

FIGURE 4.1 Examples of the difference in failure appearance between ductile and brittle materials for three basic loading mechanisms.

PHOTO 4.1 Three conveyor rollers. The center one is "as supplied," while the outer two have failed from contact stresses. The left roller was case hardened and cracked, while the right roller was unhardened and ductilly deformed.

Some important clues that can be realized from overload failures are the following:

- They clearly state that the part was grossly overstressed.
- The stress that caused the failure was applied immediately before the failure occurred.
- The direction of the fracture describes the direction of the failure load.
  - With brittle fractures, the crack is always perpendicular to the failure force.
  - With ductile failures, the direction of the elongation is the direction of the failure force.

PHOTO 4.2 Torsional overload failures of a through hardened truck axle and an inexpensive socket wrench extension. Note the difference in appearance.

Therefore, if two identical parts, one made of unhardened mild steel and the other one of hardened steel, are overloaded identically, the ductile mild-steel part will bend whereas the brittle hardened part snaps in two. The failure analyst can look at both of these parts and rapidly see:

- They were the result of loads applied immediately before the parts failed.
- The direction of the forces that caused the failures.

> On a microscopic basis, there is a huge difference in the ductile and brittle failure mechanisms. Ductile failures occur through rupture of the individual grains of metal and generally require a great deal of energy to propagate the crack. On the other hand, brittle fractures grow by separating the grains along their boundaries and usually don't take much energy to propagate. Occasionally, in the brittle fracture of a ductile material such as the motor shaft in Photo 4.11, the crack will stop growing because it just doesn't have enough energy to continue.

TEMPERATURE EFFECTS ON OVERLOAD FAILURES

For most metals, their ductility, or lack of ductility, is an inherent property and changes very slowly as temperatures increase. As mentioned earlier, in the range between −75°C and 250°C (−100°F and +500°F), temperature has almost no effect on the tensile, yield, or fatigue strength of most of the common steel alloys. But over the same range of temperatures, impact strengths can vary by more than a factor of ten. This loss of impact strength as the temperature decreases is usually called *low-temperature embrittlement,* and Figure 4.2 shows the effect of temperature on the toughness of two mild steels.

A good example of this low-temperature embrittlement phenomenon can be seen in the failure of a group of chairs from ski lifts. The chairs were typically operating at about −10°C (14°F) and were impact loaded when they swung and hit the top terminal supports. They were empty at the time of failure, and as with all failures, there were multiple causes. But one of the contributing causes was that these chairs were made of AISI 1030 carbon steel instead of the specified AISI 1020. The 1030 was about 25% stronger than the 1020, yet it was the stronger chairs that were failing. And, by looking at both the fracture faces and Figure 4.2, we can easily tell why they failed.

On the ski lift, the impact loads occurred as the chairs swung and hit the support for the wire rope wheel at the top of the ski hill. At the operating temperatures, the specified weaker material would require between 100 and 115 J (75 and 85 ft.-lb) of energy, whereas the stronger AISI 1030 chairs would fail at about one tenth of that value.

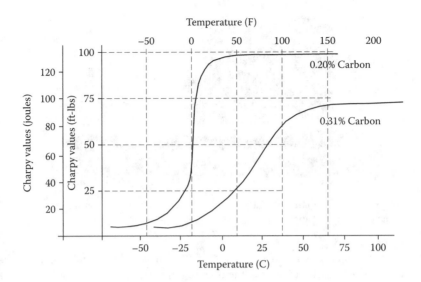

FIGURE 4.2 Comparing the impact strength of two low-carbon steels.

> The first time we ever became involved with a low-temperature embrittlement failure was in an ancient manufacturing plant. Someone had fabricated a rigging bracket from a piece of AISI 1045 steel and had welded it to a column. For years, the bracket was used every summer for lifting up to 900 kg (2000 lb) of material to the roof of a building. Then, one year, the outage was held on a cold day in November and the bracket snapped when there was almost no load on it. There is always a trade-off for higher strength.

Another important point to recognize is that this low-temperature embrittlement does not affect the common 300 series stainless steels. In that range shown in Figure 4.2, where carbon steel toughness drops precipitously, a typical 304 or 316 stainless steel part has a Charpy V-notch toughness over 135 J (100 ft.lb) that actually increases slightly as the temperature decreases.

ANALYSIS OF DUCTILE FAILURES

Some key points in looking at a ductile failure and understanding the forces involved are:

- The fact that an overload failure occurs tells that the load was greater than the yield strength of the part.
- The distortion lies in the direction of the forces that caused the failure.
- The applied fracture force was greater than any of the previous loadings.
- Microscopically, the failure occurs when the metal grains are torn apart.
- Ductile fractures usually require a great deal of energy.

The following are a series of photos that illustrate some ductile overload failures.

Photo 4.3 shows the ductile torsional overload failure of a splined shaft from a reducer. Note the twisting (ductile) distortion of the splines.

Photo 4.4 shows the ductile distortion of the end of a mild-steel motor shaft, and the end of the keyway has been seriously deformed. The distortion is at the end of the key because the stress was greater than the yield strength of the steel, but the portion of the shaft under the coupling was reinforced by the key and coupling hub fit and didn't deform.

Photo 4.5 is a ductile tension overload of a 1.3-cm (5/8-in.) bolt. The arrow points to the beginnings of the classic cup-and-cone fracture usually seen in the later stages of a pure tension failure.

PHOTO 4.3 The input spline of a small reducer. This shaft failed as a result of another failure that jammed the reducer.

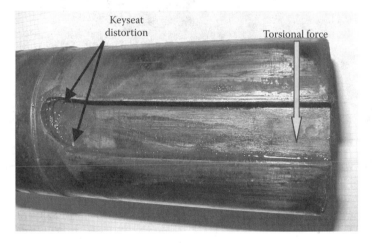

PHOTO 4.4 The keyway deformation is proof that the shaft has been overloaded.

PHOTO 4.5 The ductile overload of a 16 mm (5/8″) bolt that resulted from overtightening.

Photo 4.6 and Photo 4.7 are evidence of the ductile overload of a weldment on a piece of construction equipment. They show the right and left ends of a welded reinforcement that projects through the sides of a box beam on a crane. The cracked paint clearly shows the stress field and the areas at the ends of the welds that have been distorted. Notice that the left photo shows much

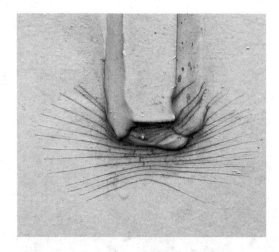

PHOTO 4.6

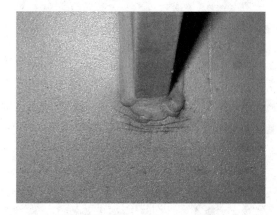

PHOTO 4.7 The cracked paint shows proof of the deformation that took place when this crane beam was overloaded.

more distortion than the other — proof that this symmetrical beam was side loaded and the load wasn't centered when it was lifted.

ANALYSIS OF BRITTLE FRACTURES

Some of the key points in a brittle fracture are the same as a ductile failure, i.e., the load is greater than the yield strength of the part, the fracture force is greater than any of the previous loadings, and the load was applied immediately before the failure took place.

Some noteworthy differences are:

- In a brittle fracture, the fracture plane is perpendicular to the direction of the forces that caused the failure.
- A brittle fracture uses much less energy than a ductile failure. Microscopically, the brittle failure happens as a fracture between the grains (along the grain boundaries), and once the crack starts, it typically does not take a lot of energy to propagate.
- The point at which the failure started can usually be seen by following the chevron marks (surface features on the fracture face) that point toward the origin.
- The crack growth is very rapid, approaching 3000 m/sec (10,000 ft/sec in steel).

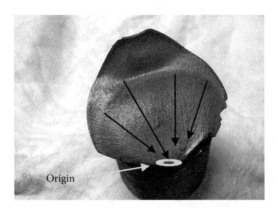

PHOTO 4.8 A reducer input shaft failure. We've put a series of lines on the surface to accent the "Chevron marks" that point to the crack origin.

PHOTO 4.9 The cracks in these two rolls run down the sides, proof that excessive hoop stress was involved.

Photo 4.8 shows the brittle fracture of a large reducer input shaft that was caused by an impact load when the reducer was bumped. The chevron marks, the lines on the fracture surface, are a great aid in the diagnosis of brittle fractures because they point to where the failure started. (We've added four arrows emphasizing the direction of the chevron marks.) An interesting point is that most persons would expect the fracture to begin at the keyway notch, but the actual origin is just to the left of it.

In Photo 4.9 can be seen the brittle tension fractures of two pulverizer rolls. Excessive hoop stress caused these fractures of extremely hard and strong, but very brittle, castings. The cracks run straight down the side of the roll, perpendicular to the hoop stress, and the rolls show no visible distortion or elongation.

Photo 4.10 shows the brittle fracture of a universal joint caused by an excessive bending (operating) load. Again, there is no visible distortion. On this fracture surface the chevron marks that point to the origin are obvious.

Chevron Marks

As shown in several of the earlier photos, a valuable feature on the fracture face of a brittle fracture is that the chevron marks point to where the failure started. They show where to take the metallurgical samples needed to check as to whether the metallurgy or condition of the material played a part in the failure.

There are times when it is difficult to see the chevron marks, and a careful inspection with a narrow-beamed light source, as shown in Figure 4.3, is extremely helpful. Experience has shown that the best interpretation of surface features can be done with a narrow, bright light beam at a very shallow angle.

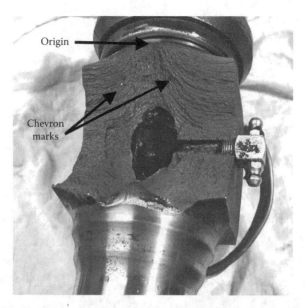

PHOTO 4.10 A universal joint with the Chevron marks clearly pointing at the origin of this brittle fracture.

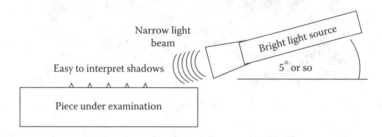

FIGURE 4.3

Figure 4.3 shows how you should use the light to accentuate the surface features and the two photos, Photo 4.11 and Photo 4.12, demonstrate the actual effects. These two pictures 4.11 and 4.12 are of the brittle fracture face from a crane hook and the only thing that changed between them was the angle of the light beam.

PHOTO 4.11 Easy to identify failure origin because the low light angle accentuated the Chevron marks.

PHOTO 4.12 This is the same piece as shown in Photo 4.11, but the direct lighting has hidden the surface features.

UNUSUAL CONDITIONS

BRITTLE FRACTURES OF DUCTILE MATERIALS

Rarely, perhaps once in every hundred industrial failures, there will be a brittle fracture of a ductile material. Every history buff knows the story of the Liberty ships in World War II and their problems with brittle fractures. Several of these ships were lost in the North Atlantic with their destruction initially being blamed on German submarines. Even though the material of construction should have been ductile, many of these Liberty ships suffered major fractures, essentially without warning, and one even sank in the harbor before being loaded. This outbreak of brittle fractures resulted in several detailed studies of the causes and a much better understanding of the problem.

Brittle fractures of ductile materials essentially take place under one of the following conditions:

1. The part is loaded so highly and rapidly that the ductile material doesn't have time to deform before it is fractured.
2. The ductile material is so constrained that it can't deform before it fractures.

Either the ductile materials don't have enough time to behave in a ductile manner or the design of the structure is such that the material can't deform in a ductile manner, and the result is that a brittle fracture occurs.

> Don't confuse this with the low-temperature embrittlement of ductile materials. In low-temperature embrittlement, the piece fails because it is impact loaded but doesn't have the toughness to resist catastrophic failure. In these failures, the results are the same but the parts are not impact loaded.

Rapid Force Application

Two examples of brittle fractures of ductile materials that were caused by the rapid force application are:

1. In an analysis we worked on, a cable-supported industrial elevator mechanism fell about 30 m (100 ft). When the carriage reached the end of the cable, traveling at an incredible speed, the cable snapped like a piece of string and, because the load was applied so rapidly, the ductile individual wires fractured in a brittle manner.
2. Photo 4.13 shows the coupling end of a 133 kw (100-hp), 3600 r/min mixer motor shaft with a diagonal crack in it. The mixer motor was running at full speed when some product was accidentally dropped into the machine, immediately jamming it. An analysis showed the shaft was AISI 1035 steel, a relatively ductile material, which according to the chart

PHOTO 4.13 The end of a 3600 rpm, 133 kw (100 hp) motor shaft that stopped instantly.

in Chapter 3, should elongate by about 18% before failure. Nevertheless, one can easily see a diagonal crack that appears to be the torsional failure of a brittle material. However, further examination of the parts and the site shows:

a. The fracture is not on a 45° angle as would be expected from a true brittle fracture.
b. There is some deformation at the end of the keyway.
c. The crack stopped halfway across the shaft.

Looking at these apparent incongruities, (a) and (b) happened because there was some ductility to the shaft, and it had enough time to deflect a little, but not what would have normally been expected. (If the shaft had reacted in a purely ductile manner, the end of it would have been turned off and the fracture would have been essentially straight across the shaft, like the inexpensive socket extension in Photo 4.2.) Item (c), shows that there wasn't enough energy to drive the crack all the way across the shaft because the failure was converting from cleavage to ductile rupture.

Constrained Materials

The other situation where ductile materials act in a brittle manner is when they are so constrained that they can't deform. (This was the basic cause of the Liberty ship failures.) For example, if there is a square corner weld, where several heavy plates are welded together with significant residual stress, the weld material cannot deform to release the stress in a ductile manner because it is being pulled in three directions at once. The result is a fracture that often occurs at stresses below the yield point.

An example of these failures occurred in a large welded structure. There were several 1- and 1.25-cm (⅜-in. and ½-in.) sections welded together as reinforcement for a holding tank. The reinforcing steel was overstressed. One of the 1.25-cm-thick (½-in.) pieces was about 45-cm (18-in.) long. Immediately adjacent to the point where it was welded to two other sections, a 10-cm-long (4-in.) section of the plate necked down (ductile deformation) until it was only 1-cm (0.40-in.) thick. Then it suddenly fractured in a brittle manner. Along the 10-cm (4-in.) section, the crack looked like ductile elongation but the remaining 35 cm (14 in.) fractured as a brittle overload.

Photo 4.14 shows the brittle fracture of a highly constrained weld. (The small black spheres in the photo are slag from the process operations.) This is a weld on a lifting beam from a steel mill. The weld material has a catalog elongation of about 25%, yet it fractured in a brittle manner because it was very heavily stressed, and it couldn't deform in a ductile manner because it was a 0.6-cm (¼-in.) throat tack weld on a 2.5-cm-thick (1-in.) plate.

PHOTO 4.14 The brittle fracture through the throat of a small weld on a heavy section.

Notch Sensitivity of Brittle Materials

In the chapter on fatigue failures, there is a lengthy discussion on stress concentrations, how they increase the stress in their immediate vicinity, and how they contribute to fatigue failures. In a similar manner, brittle materials are very susceptible to sharp corners and notches. If the notch is severe enough and the stress high enough, the part will fracture even though the stress may nominally be only 75% of the yield (or tensile) strength. Here we show two examples of this.

Photo 4.15 and Photo 4.16 show two views of a hardened cast alloy hammer from an 1100-t/h crusher used at a Canadian mine site. This very hard iron alloy is designed to be abrasion resistant, but it is relatively notch sensitive. The notch, seen at the arrows in both photos, was caused by an error in the manufacturing practice.

Photo 4.15 shows the new hammer assembly in position after it has shattered, pretty much destroying everything in sight. The second photo is a close-up of the left side of the hammer showing an obvious brittle fracture with the chevron marks pointing toward two small black areas. These are casting defects, which resulted in the notches that substantially weakened the part.

In the aforementioned example, the load was applied very rapidly, but for a notch to cause these problems doesn't require impact loading. Photo 4.17 shows a fork from a 36-t (40-t) capacity lift truck that was slowly carrying a heavy load across some uneven ground when the fork snapped. The notches that caused the failure can readily be seen, and it is obvious from their dark color that

PHOTO 4.15 An overall view of the broken hammer. This brittle fracture started at the arrow.

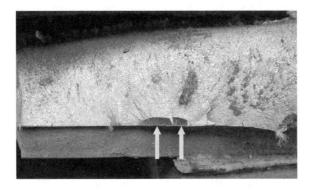

PHOTO 4.16 A close up view of the fracture origins showing the block, oxide-coated casting defect that hastened the failure.

PHOTO 4.17 This brittle fracture of a left truck fork started at the old cracks (notches) in the center.

they had been there for some time. Two observations on the load application and subsequent failure are as follows:

- The photo was taken in the field, and it had been raining.
- The truck was not overloaded, and the load was properly positioned.

SUMMARY

Overloading a part can result in either a ductile or a brittle fracture, depending on the material characteristics and the rate of load application. Most of the materials we deal with on a regular basis are ductile.

There is a tremendous difference between tensile strength and impact strength. The impact strength, i.e., toughness, for many steels changes significantly between –15°C and 50°C (0°F and 125°F), whereas the tensile and yield strengths are essentially unchanged.

Under certain conditions, i.e., when loads are applied very rapidly or the piece is very highly constrained, ductile materials can behave as though they were brittle. But brittle materials behave in a ductile manner only at highly elevated temperatures.

Generally, the harder and stronger a material is, the more sensitive it is to notches.

5 Fatigue Failures (Part 1): The Basics

Fatigue causes more mechanical failures than any other mechanism. It caused my bike spokes to break and the bearings in our saw to rumble. It certainly varies with the type of machinery, however, it causes at least 80% (and possibly 90%) of all mechanical failures.

By definition, overload failures take place during one force application and at loads equal to or greater than the yield strength of a material. On the other hand, fatigue failures generally occur at loads less than that needed to cause deformation, and they usually require many stress cycles to initiate and then grow across a part.

When an overload failure occurs, inspection of the failure tells the direction of the applied force, the rate at which the force was applied, and the fact that the load that caused the failure was applied in the millisecond or so before the part failed. In contrast, the event that started a high-cycle fatigue failure may have been initially applied one hundred thousand, a million, or even twenty million or more stress cycles in the past. Fortunately, careful inspection of the part, and the fracture face, in particular, can show the relative load, direction, and uniformity of the load application, whether stress concentrations played an important part, and many other important facts about the causes.

FATIGUE FAILURE CATEGORIES

There aren't any formal definitions for the categories of fatigue failures, but three classifications in general use are:

1. High-cycle fatigue — Where it takes more than 10,000 stress cycles to fail, it is called *high-cycle fatigue*. These probably account for more than 90% of all fatigue failures. They are much more common than either of the other categories, and this chapter will cover them in detail. Most high-cycle fatigue failures take more than 100,000 cycles, and some take many, many million cycles from initial force application to final failure.
2. Low-cycle fatigue — When the failure takes fewer than 10,000 but more than about 25 stress cycles from the application of the force to final failure, it is considered a *low-cycle fatigue* failure. They are generally not very common in industrial equipment and, referring to the S–N curve in Figure 5.1, one can see they involve higher stresses than high-cycle failures.
3. Very-low-cycle fatigue (VLCF) — One interesting point about low- and high-cycle fatigue failures is that they show no visible plastic deformation when the part fractures and the pieces are examined. For example, there may be some deformation of the last thread on a failed bolt but there is no visible deformation of the body of the part. However, VLCF failures:
   a. Take only a few stress applications, typically less than 10, from initiation to final failure
   b. Show some visible deformation in ductile materials
   c. Indicate that the applied force was only slightly less than what would cause an overload failure
   d. Are not very common in industrial equipment but are seen more often than low-cycle fatigue examples

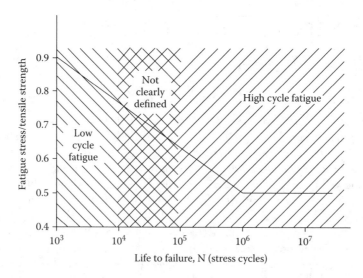

FIGURE 5.1 An S/N chart showing the ranges for low and high cycle fatigue failures. (VLCF would be off the left side of the chart.)

Fatigue failures cannot occur from pure compressive stresses. For fatigue to happen, the part has to undergo periodic tensile stress and then relaxation, and this cycle has to be repeated many times. How long does it actually take for a fatigue crack to grow across a shaft or bolt? Some very heavily loaded situations have only taken a few load applications, whereas others have slowly progressed for years. However some insights about them are:

- Once a fatigue crack gets big enough to be easily seen by the naked eye (more than 0.5 mm [0.020 in.] or so), it always grows perpendicular to the plane of maximum stress. This characteristic is invaluable for its ability to point to the source of the load that caused the failure.
- Most fatigue failures we've seen on components such as pumps and motor shafts have taken less than 10,000,000 total cycles from start-up to final failure, but there have been some that have taken much longer.
- Under light and variable loads, especially when corrosion is present, it is not unusual to see cracks take 40 to 50 million cycles to propagate across a part. (Corrosion plays this significant role because it continuously reduces the fatigue strength of the material and, if a cyclical load is present, a crack will eventually start even under very light loads. This results in a situation where there is a long incubation before the crack starts and then an extremely slow-growing fracture, where the crack propagation per stress cycle is almost infinitesimal.)
- Stress or load variations often complicate the diagnosis of fatigue failures. For example, in a machine that sees varying loads, such as a shaft that may turn at 100 r/min, the thought is that it sees 144,000 fatigue cycles per day. But if it is heavily loaded for 10% of the time and the rest of the time the load is very light, it actually sees 14,400 significant stress cycles per day.

Before becoming involved with the analysis of fatigue fractures, it would be a good idea to better understand the following:

- The concept of stress concentrations because of their importance in a very high percentage of fatigue failures
- The effect that metallurgy has on fatigue crack origins

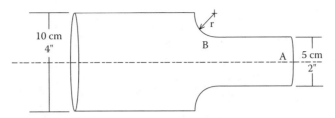

FIGURE 5.2 A cross-section view of a shaft subjected to fatigue loads.

STRESS CONCENTRATIONS

Stress concentrations, or *stress risers* as they are frequently known, are features of the part that multiply the local stress. In effect, because of a change in shape or a change in metallurgy, the effective stress in the immediate area of the stress concentrator is greatly increased. Figure 5.2 shows a shaft, and the following chart lists the ranges of stress concentration factors that are caused by the change in shape. These higher stresses occur on the shaft surface at the very top of point B and, in an engineering analysis of the shaft, the field stress at point A would be multiplied by the stress concentration factor to obtain the actual stress at point B.

Stress Concentration Factors for Figure 5.2

Radius (mm/in.)	Tension	Torsion	Bending
0.4 mm/0.015	8	2.5	5.0
1.5 mm/0.60	3.7	2.0	2.7
3.0 mm/0.12	2.3	1.6	2.2
6.2 mm/0.25	1.8	1.4	1.7

Source: Data from Peterson's Stress Concentration Factors, Walter Pilkey, Ed., John Wiley & Sons, New York, 1997 .

Commenting on the stress concentration factors for Figure 5.2:

1. The symbol normally used for the stress concentration factor is K_t.
2. There are substantial differences in stress concentration factors for tension, torsion, and bending loads. This is because the stressed volume of the shaft changes, depending on the way the stress is applied. For example, in tension, the entire shaft cross section at B is equally stressed. However, if torsion were to be applied to the piece, only the outermost fibers would see the peak stresses.
3. Stress concentration factors can easily be calculated by most FEA programs. However, the most complete text on them was probably done by R.E. Peterson, a Westinghouse engineer, who spent a lifetime compiling invaluable data on stress concentration factors for various component geometries.
4. An interesting point to realize is that the effective stress concentration factor at the tip of a crack will vary depending on the crack length and properties of the material, but we have seen data that states that in some common materials it can be as high as 25.

Some of the most common locations for stress concentration problems include:

- Steps or grooves in shafts
- Welds in the stressed area of a component
- Holes in components

- Keyways and key seats
- On a bolt body, the transition to the threaded section
- Shrink-fitted components with sharp corners
- Rough surfaces perpendicular to the stress field

STRUCTURE CHANGES CAUSED BY HIGH-CYCLE FATIGUE

Continuing our quest to better understand how and why high-cycle-fatigue failures occur, a basic understanding of the component metallurgy is helpful. Figure 5.3 is a sketch that shows the surface of a metal piece and its atomic structure. Ideally, it would be a single perfect crystal with a uniform structure extending to the edges of the part but, in reality, there are many crystals, and inside each are a number of grains. Unfortunately, within the grains are a tremendous number of dislocations, and these irregularities are weak points in the structure.

As a metal is repeatedly stressed, the dislocations gradually work their way through the structure and begin to align themselves at grain boundaries. The more heavily the material is stressed, the more rapidly the dislocations move until they eventually line up to create a weakness. With continued cyclical stressing, this weak point becomes a tiny crack, which eventually propagates across the structure.

As the crack grows across the part, it starts very slowly but becomes progressively faster until the piece breaks in two. Figure 5.4 is the sketch of a basic fatigue failure face showing a crack that started at the origin and grew slowly across the fatigue zone until it reached the instantaneous (or fast fracture) zone. At that point, the crack growth rate instantly changed from a thousandth of an inch or so per stress cycle to thousands of feet per second and the piece fractures.

DIAGNOSING A HIGH-CYCLE-FATIGUE FAILURE

In the following subsection most of the drawings and photos are of shaft failures, but the diagnostic techniques we mention can be used for all fatigue failures.

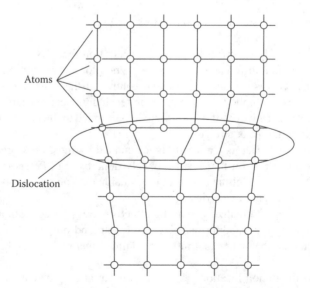

FIGURE 5.3 The cross section of a metal crystal showing a dislocation in the structure.

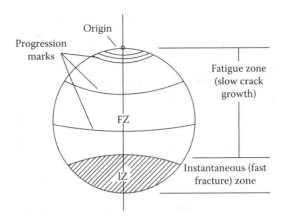

FIGURE 5.4 A view of a basic plane bending fatigue failure.

In analyzing the fracture face in Figure 5.4 and defining the areas and features, one can see the following:

- Origin — This is the point where the cracking actually started and is the oldest and smoothest part of the fracture face.
- From there, the crack slowly grew across the fatigue zone, advancing a miniscule distance with each stress cycle. Because of the very slow crack growth, this area of the fracture face is relatively smooth, and one clue to the age of the crack is the smoothness of the fatigue zone. Consideration has to be given to the grain size of the material, but the smoother the surface, the slower the growth and older the crack.
- If the cyclical load on the part isn't constant while the crack is growing, the growth rate and the surface appearance will change, and the results of these load changes are the progression marks. They are a calendar detailing the changes in loading over the life of the crack. They are also called *beach marks* or *stop marks*, but the feature describes the growth of the crack across the piece, and our opinion is that *progression marks* is a much more accurate and descriptive term.
- When the load on the piece becomes greater than the remaining strength, the piece suddenly fractures across the instantaneous (or fast fracture) zone (IZ). In this final fracture, the failure may be ductile, but in most instances (probably more than 95% of the time in industrial equipment), the instantaneous zone is a brittle fracture, and the surface is rough and crystalline in appearance. A valuable point is that the size of the instantaneous zone is an indication of the stress on the part at the time of the final fracture.

PROGRESSION MARK BASICS

Progression marks tell us how the crack grew and they are an incredibly valuable tool to help in understanding what happened during the life of the failure. The next few paragraphs will introduce the basics, but we will revisit the subject of progression marks several times in this chapter.

To understand how a progression mark is developed, visualize a crack growing across a part. It grows a tiny bit with each stress cycle, but if the magnitude of this stress decreases, the crack growth essentially stops, even though the part is still being cyclically loaded. Later, when the load increases again, the crack resumes growing. This period of no growth results in the change in surface appearance that is called a *progression mark*.

Fatigue failures are caused by cyclical stresses, but if there is no change in the range of those stresses, there won't be any progression marks. For example, if the ammeter on a pump motor indicates that the load is steady with no changes, there would be no drastic changes in the crack growth rate and no progression marks on the fracture face. But if the ammeter (and the stress level) was varying widely, the crack would alternately grow fast and slow, and there would be series of progression marks that would show an account of the number of variations.

There are times when these marks can be traced back to specific events in a machine's operating history, giving a better understanding of how and when the failure developed. In one example, we looked at a 25-cm [10-in.] compressor shaft that had three very definite progression marks over about 0.5 cm [1/4 in.], while the rest of the failure face was extremely smooth. The owner was able to trace these progression marks back to one of the roots of the failure being a process upset that put the unit into a surge condition.

Furthermore, when analyzing the fracture forces and crack growth, knowledge of fracture mechanics is definitely helpful. For example, in high-cycle-fatigue applications, progression marks result from variations in the plastic zone at the crack tip during changes in field stress and, if the crack toughness of the material is known, the exact stress at the time of final fracture can be calculated from the size of the IZ.

In addition to the process mentioned earlier, there is actually a second mechanism that generates progression marks. Depending on the stress on the shaft and the material properties, in the latter stages of a fracture's life (possibly the last hundred or so stress cycles) there are times when the crack grows a visible distance with each stress cycle, resulting in a series of progression marks immediately before the final failure, as shown in Photo 5.1. This is a 10-cm (4-in.) shaft from a trolley drive in a steel mill. A careful analysis showed the cracking started in a fretted area adjacent to the key. It slowly grew across the fracture face and then the progression marks began approximately at the arrows. These progression marks are also fatigue striations and show the crack growth in the last few cycles. (In addition, the small IZ shows that the shaft was very lightly loaded at the time of final failure.)

Occasionally, there is confusion between progression marks and fatigue striations. Fatigue striations show each stress cycle experienced by the part and, for the portion of the failure up to the tips of the arrows, are usually visible only under high magnification, whereas progression marks are visible to the naked eye. Fatigue striations, as shown in Figure 5.5, lie between the progression

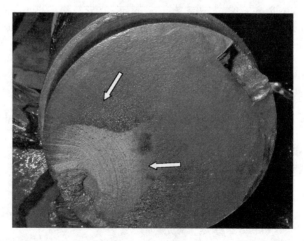

PHOTO 5.1 This fatigue failure started at the keyway. The fracture face is relatively smooth up until the ends of the two arrows. From there to the final fracture progression marks are readily visible.

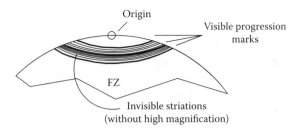

FIGURE 5.5 Progression marks are readily visible features on the fracture surface, while striations are generally visible under magnification.

marks. In some alloys such as aluminum, they are relatively easy to see whereas in others, such as most of the stainless steel alloys, they are almost impossible to find even with an electron microscope. (We should repeat that the progression marks referred to in the previous paragraph, the ones that occur in the last hundred or so cycles immediately before the final failure should also be considered as fatigue striations.)

If there is a real need to understand how long it took between fracture initiation and final failure, a common technique is to look at the fracture face using relatively high-powered electron microscope, usually between 500 and 3000 X, and slowly scanning the fracture surface. When fatigue striations are found in several areas, the crack growth rate across those areas is calculated and integrated across the entire fracture surface, allowing a reasonably accurate approximation of the growth rate and time from initiation to final failure.

FRACTURE GROWTH AND UNDERSTANDING THE SOURCE OF THE STRESS — ROTATING BENDING VS. PLANE BENDING

Fortunately from a diagnostic viewpoint, there is a substantial difference in appearance between a plane bending failure and a rotating bending failure. In a plane-bending failure, such as would occur with one-way bending of a leaf spring or a diving board, the crack grows straight across the part as shown in Figure 5.4. However, if there is a rotating bending load, the crack growth would be asymmetrical, and the fracture face would look like that shown in Figure 5.6.

An example of fatigue cycling caused by a rotating bending load could be a motor shaft with a belt load on it or a reducer with a chain drive. Visualizing the shaft between the load and the first bearing, we see that as the shaft rotates, the stress at every point on the shaft surface varies, first being in compression, then half a rotation later, in tension. This rotating cyclical tension and compression, results in the dislocation migration that contributes to the fatigue crack initiation.

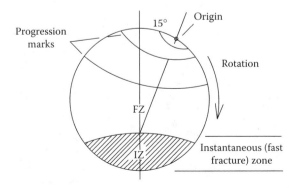

FIGURE 5.6 A rotating bending failure with the bisector of the instantaneous zone not pointing at the origin.

A comparison between these two fracture faces shows that:

1. In the plane-bending failure, Figure 5.4, the bisector of the IZ points to the crack origin. But in the rotating-bending failure (Figure 5.6), because of the changes in stress distribution as the shaft rotates, the crack grows unequally and the bisector does not point to the origin. (This angle is shown as 15° but will vary depending on the relative rotating- and plane-bending loads. For example, if there is a large plane-bending component of the fatigue load and a small rotating-bending component, the angle will be close to zero.)
2. As a result of the unequal crack growth, it is possible to tell the direction of component rotation.
3. The progression marks, if any, will mirror the crack growth and will grow unequally.

There are an almost infinite number of variations in the appearance of fatigue failures. But we'll start with a couple of basic examples and gradually introduce more confusion into the diagnostic procedures.

Below are two photos of failure faces. Photo 5.2 shows a portion of the fatigue face that is essentially a plane-bending failure. The origin is fairly obvious as are the progression marks that appear to grow straight across the shaft. Also, the change in surface roughness is one of the important tools for identifying that it is a fatigue failure, i.e., as the crack grew across the shaft, the surface became progressively rougher because of the increased rate of crack growth.

> Actually, this was a rotating agitator shaft, and a careful inspection shows that the progression marks are growing very slightly more on one side than on the other, because there was a small component of rotating-bending stress affecting the shaft.

Photo 5.3 is a view of a rotating-bending failure. Inspection shows that:

1. By following the progression marks backward, one can see that the cracking started in the upper-left corner of the keyway.
2. The progression marks grew unequally across the shaft, and the bisector of the IZ does not point at the origin.
3. The shaft was rotating in a clockwise direction.
4. The many progression marks show that there were a large number of changes in the shaft stress.

We say the shaft was rotating in a clockwise manner and it is difficult to put into words why the crack grows unequally, but it is easy to visualize with a simple experiment. Take a rolled tube

PHOTO 5.2 A fatigue failure with an obvious origin and a series of well-defined progression marks.

PHOTO 5.3 The progression marks show this failure started at the upper left corner of the keyway, but the sector of the instantaneous zone doesn't point to the origin, showing the cause was rotating bending.

or something similar and rotate it in your hands in the same manner the shaft would rotate in operation. Next, mark a crack in the center of the shaft section and perpendicular to the long axis. Then visualize the bending stress in the cracked shaft as it rotates. Note how, as you rotate the shaft and the simulated crack rotates into view, the crack would have little effect on the stress in the half of the shaft you can see. But as the shaft continues to rotate, the effect grows substantially. The same experiment could be conducted with strain gages to show precisely how the rotationally loaded shaft crack grows unequally.

Always look at the failure, and look for progression marks or other clues to the shaft rotation. Centrifugal pumps will pump and squirrel cage blowers will blow air even if they are rotating in the wrong direction — and the chance of connecting an electric motor with the correct rotation is 50%. We have seen many failures where the operators or maintenance personnel thought it was running in the right direction but never carefully checked. In one case, a plant had gone through 5 years of performance problems with an air-handling system. Then we were asked to balance the fan to reduce the vibration and, when our technician said it was running in the wrong direction, they said, "It couldn't be! Nobody's worked on it for 5 years!"

Continuing support on this need for analyzing the direction of rotation ... A good friend of ours, who was a large motor consultant for an electrical equipment manufacturer, once said that over 40% of the motor failures that he had seen (he defined these large motors as those with separate cooling fans) had their squirrel cage fans running in the wrong direction. Wrong direction of rotation results in less cooling air. Reduced cooling air results in higher winding temperatures and, according to Arrhenius' rule, more rapid insulation degradation and shortened motor life.

PROGRESSION MARKS AND VARYING STRESS LEVELS

An important point is that the progression marks should always be examined to understand how the fracture forces changed during operation. As shown earlier, in a plane-bending failure, the progression marks will grow essentially uniformly across the piece, whereas a rotating bending failure shows asymmetrical growth. If the forces on the part change during operation, the shape of the progression marks will also change. Figure 5.7 shows a shaft with irregular progression marks,

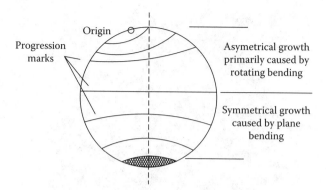

FIGURE 5.7 In this example the change in the progression mark growth pattern occurred because the forces on the piece changed.

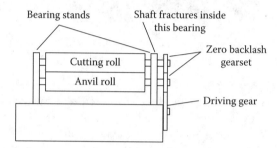

FIGURE 5.8 The fracture face in Figure 5.7 showed that the major force causing the fracture was rotating bending, which only could have come from gears.

i.e., they first grow asymmetrically, then they grow straight across the shaft. Inspection of the failure face shows the cracking began primarily because of rotating bending, and then, as the crack progressed across the shaft, the operating characteristics changed, and the driving force behind the crack growth was plane bending.

The importance of carefully inspecting and understanding the progression marks may be difficult to understand, but the following example will illustrate it. The fracture in Figure 5.7 was actually one of a series of a half dozen failures from a group of very similar machines. Analysis of the six fractures showed almost identical characteristics. The bisectors of the IZ did not point at the origins, and in the early stages of the failures the progression marks were growing eccentrically. Yet, later on, as the fracture reached the halfway point on the shaft, the progression marks began to grow straight across the shaft face.

The shaft was on a cutting machine, and a sketch of the machine is shown in Figure 5.8. It was used to precisely trim a consumer product and had a blade that struck the anvil roll once per revolution. The plant engineers and supervision were positive that incorrect parts on the cutting roll caused the failure, and they went over and over the very detailed setup procedures with their personnel. But the setting of the zero backlash gears was left to the operators. Inspection of the progression marks in Figure 5.7 tells us that the forces that started the failure were the rotating bending fatigue forces that could only have come from the gears, and the plane bending forces didn't come into significant play until half of the shaft was gone.

PROGRESSION MARKS AND STRESS CONCENTRATIONS

A sometimes confusing feature of progression marks comes when stress concentrations are present in a plane bending failure. Given in Figure 5.9 are the sections of two shafts with examples of progression marks. Both shafts show progression marks that are concave, and the difference is that

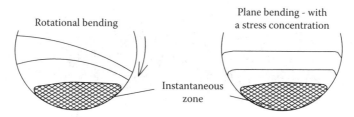

FIGURE 5.9 Note how the progression marks leading up to the instantaneous zone differ between rotating and plane bending failures with high stress concentrations.

the rotation bending progression marks are eccentric to the bisector, whereas the plane-bending failure has marks that are essentially symmetrical. The reason the plane-bending progression marks turn downward at the edges is that the high stress concentration (and the resultant increase in local stress) causes the cracks to grow much faster at the edges.

Two places where this type of plane-bending progression mark is common are fasteners (with a high stress concentration at the first thread) and shafts that have a sharp groove machined in them.

RATCHET MARKS

If a piece is repeatedly stressed to a level just a little above the fatigue strength, it will eventually develop a fatigue crack and fail. However, because it was lightly loaded, like the pieces in Photo 5.3 and Figure 5.7, the cracking will only start in one place on the exterior of the piece, at a point where some minor weakness exists. However, if the part were more heavily stressed, cracks will start in several places almost simultaneously, as shown in Photo 5.4, where the arrows point to three separate crack origins. As the effective stress increases, there are more places where the local fatigue strength is exceeded at the same time.

Looking at this photo, one can see that in between these three origins are two ratchet marks. The three cracks don't start on precisely the same plane along the axis of the shaft, and the ratchet marks are essentially boundaries between the fracture planes. As the cracks grow inward and the fracture planes unite, the ratchet marks disappear.

The presence of many ratchet marks is usually an indication of high stress concentrations. For example, the setscrew marks in Photo 5.4 are the stress concentrations that caused the fractures to start at those points.

> However, beware of red herrings. The setscrew marks in the shaft had almost nothing to do with the failure. The black mark to the left side of the shaft is a scarf mark from burning a bearing off the shaft. When the bearing was burned off, the unequal heating of the shaft caused it to permanently bend, and this resulted in fatigue cycling of the shaft as it rotated in operation. I won't say it is impossible to cut a bearing off a shaft with a torch and not bend the shaft; it's just that, despite many attempts, I have never seen it done. Cutting bearings off a shaft with a torch is a common technique in many plants, and the only way to accurately tell if the shaft has been bent is to check the runout before and after cutting.

Most fatigue cracks have some readily visible ratchet marks. In some cases, such as when they appear after the crack has been well established, they don't tell us very much and can be ignored, but there are other times when we can learn a great deal from them. Two examples of this are:

- Frequently, the ratchet mark growth direction is important. When two ratchet marks grow in different directions, the primary origin lies between them. Photo 5.5 shows a gear

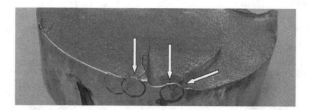

PHOTO 5.4 There are failure origins at the tips of the three arrows. Between the origins are ratchet marks, a result of the three crack origins being on slightly different planes.

PHOTO 5.5 The fact that the ratchet marks on this fatigue fracture face grow in different directions shows the origin was between them.

PHOTO 5.6 These multiple ratchet marks result from many crack origins — either from high stress and/or from high stress concentrations.

> tooth. The roughness of the surface shows that the crack grew rapidly and that it is a low-cycle fatigue failure. Further inspection shows that the ratchet marks in the center of the tooth grew in opposite directions, and the primary origin lies between them along the upper edge of the tooth at the arrow. If a detailed metallurgical analysis were to be performed, it would be at this point.

- When a failure has a great many ratchet marks, suspect high stress concentrations.

An example of a high stress concentration and many ratchet marks can be seen in Photo 5.6 showing a small portion of the fatigue failure face of a 45-cm-diameter (18-in.) kiln trunnion roller shaft. The face has been wire-brushed clean, and the many ratchet marks are readily visible. They show that there were numerous failure origins and are indicative of either a very high operating stress or high stress concentrations. However:

- These start just below a layer on the surface, showing that the shaft material varies, with a different layer on the shaft surface.
- There are no progression marks — indicating that the stress on the shaft was constant. There were some gouges and other surface damage that can be seen in the photo.
- The IZ was very small, less than 13 cm (5 in.) across and less than 7% of the total area, showing that the stress on the shaft at the time of failure was low.

Why did the shaft fail? It had been built up by welding, without concern or an understanding of the loads, the stress concentrations, the strength of the weld, or the residual stresses.

ROTATING-BENDING FAILURES WITH MULTIPLE ORIGINS

As mentioned earlier, when a part, such as a motor shaft, is subjected to a rotating-bending fatigue stress and the stress is more than the fatigue strength, it eventually fails. If the load is relatively light, the fatigue strength is exceeded at only one weak point, cracking starts at only one place, and there is a single origin as shown in Figure 5.6.

However, if that same rotating piece were loaded more heavily, the cracking would start at many points around the perimeter, and the appearance changes drastically, as can be seen in Photo 5.7. This shows a rotating-bending fatigue failure where cracking has started around the entire perimeter. The crack grew inward and left a relatively small IZ. This photo is a good example of several features of fatigue crack analysis in that:

- There are no progression marks, showing that the load didn't vary during the life of the crack.
- The IZ is small, indicating that the load was relatively light.
- In view of the fact that the load is light and unvarying, the question is "Why did the piece crack?" An inspection of the part shows a sharp radius at a step in the shaft causing a high stress concentration. With the actual stress being the load stress multiplied by the stress concentration, the effective stress in the corner radius was enough to cause many crack origins.

A careful observer will also notice that the shaft in Photo 5.6 also shows evidence of a weld repair. However, the major problem was a razor-sharp radius at the step in the shaft.

PHOTO 5.7 A rotating bending failure with multiple origins around the perimeter. The small IZ indicates the shaft was lightly loaded at the time of final failure.

PHOTO 5.8 This motor shaft failed after 24 hours in use.

PHOTO 5.9 This one lasted only 12 hours.

DISTINGUISHING BETWEEN THE EFFECTS FROM STRESS AND STRESS CONCENTRATIONS

The two photos, Photo 5.8 and Photo 5.9, show two shafts that came off the same 900-r/min motor installation. The first one ran for 24 h, whereas the second one lasted only 12 h. Comparing them, we can see the difference between the field stress level on the shaft and the effect of stress concentrations.

Looking at the piece in Photo 5.8 (and ignoring the smearing):

- The IZ is very small, showing there was a light load at the time of failure.
- There are no progression marks, showing the loading was constant.
- There are a huge number of tiny ratchet marks, with a fracture origin between each pair.
- The fact that there were a great many fracture origins with a relatively light load indicates there is a very high stress concentration.

Next, looking at the one in Photo 5.9, we see:

- The IZ is much larger, indicating the load was much greater at the time of failure.
- Again, there are no progression marks, showing a constant load.
- There are fewer ratchet marks, therefore there are fewer fracture origins.

Comparing the two, we see the shaft in Photo 5.9 has the higher load, but less in the way of a stress concentration. If we had the two shafts in our hands, you could see that the radius at the shaft step on the first failure was as close to zero as possible, whereas the second failure has a radius of about 1 mm (0.04 in.). Unfortunately, the load on the second shaft was tremendous.

There is no precise formula, but you can see that if a shaft is lightly loaded and there are many crack origins, there has to be a high stress concentration for the fatigue strength to be simultaneously exceeded in many areas. Conversely, if the load is light and there are only one or two crack origins, the stress concentration factor must be very small.

FRACTURE FACE CONTOUR AND STRESS CONCENTRATIONS

Another clue to the severity of the stress concentration is the shape of the fracture face. Keeping in mind that the crack always grows perpendicular to the plane of maximum stress, look at Figure 5.10. Notice that the crack is concave and the steepness of the fracture curvature, the cuplike shape, and the angles at A and B, indicate that the stresses have been intensified at the corner and there is a substantial stress concentration.

As a comparison, in a component where the stress concentration is much less, the amount of concavity would also be much less. For example, the shaft section shown in Photo 5.10 has a fatigue crack that has progressed almost all the way through. Looking at the shape of the crack, it is much flatter than that shown in Figure 5.10 because the stress concentration is much less. The calculated K_t for the shaft in the photograph is 2.1.

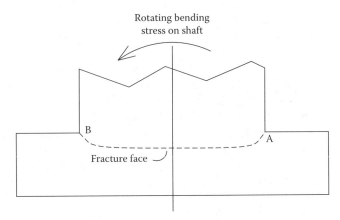

FIGURE 5.10 The contour of the fracture is always perpendicular to the plane of maximum stress. This fracture face profile suggests a very high stress concentration.

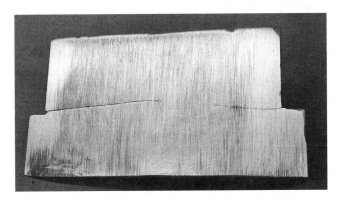

PHOTO 5.10 The calculated stress concentration for this motor shaft is 2.1.

INTERPRETING THE INSTANTANEOUS ZONE SHAPE

The IZ is the last piece of the part to fail, and the shape can show us the forces that were actually acting on the piece immediately before the final fracture. The size of the IZ gives us an idea of the forces at the time of failure and, if there are no progression marks, it is also a good indicator of initiating forces. Furthermore, in Figure 5.12 and Figure 5.13, for the various failure categories note:

- Plane-bending and reversed-bending failures — the shape of the IZ boundary will generally be convex in the last half of the failure. The presence of sharp changes near the outer surface of the piece is an indication of high stress concentrations.
- Rotational bending — Generally, the higher the total stress, the better centered the IZ.
- Multiple causes — A pure rotating load will result in a round IZ. The greater the elongation, the greater the proportion of bending load as a cause.

GUIDES TO INTERPRETING THE FATIGUE FRACTURE FACE

The first chart, Figure 5.11, is a general guide to the interpretation of fatigue failures. Using the chart and the two other guides, Figure 5.12 and Figure 5.13, the diagnosis of most fatigue failures should be greatly simplified.

SOLVING A FAILURE

Applying those guides, we frequently find that there will much more distinct ratchet marks than those seen in these three figures. An inspection of Photo 5.11 shows some of the features that are seen in attempting to solve field problems. Looking at this stainless steel shaft, we can see:

1. There are definite smooth and rough areas, the FZ and IZ, indicating a fatigue failure. (The change in surface roughness indicates that the crack grew at different speeds and proves that the failure didn't happen in a very short time.)
2. The FZ surrounds the IZ, indicating a rotating-bending failure.
3. The smoothest parts of the failure FZ are on either side of the keyway, showing that this is where the failure started.
4. The keyway is right up against a step in the shaft, multiplying the stress concentrations and helping the shaft to fail.
5. The IZ is relatively small, showing that the load at the time of final failure was not excessive. Not only do we see that the final load was low, but we also know from the lack of progression marks that the load did not change over time.
6. There are several substantial ratchet marks at the arrows. They generally indicate elevated stress concentrations, and this is supported by the small IZ. There are also several ratchet marks in the lower half of the fracture face but they are not of real concern or value because they happened long after the cracking started.
7. The IZ is oval, not round. This tells us that a substantial component of the force that caused the failure was from plane bending.

In summary, after a few minutes of analysis of the shaft in Photo 5.11, we know the shaft failed over time and was not the victim of an instantaneous overload. We also see that it was not heavily loaded, but the design was weak because the keyway was in the wrong position and the radius was not only small but also poorly machined. From this we could, in a few minutes, change the design drawing and prevent similar failures in the future. However, in analyzing the design, we should also look at the drawing to see why the plane bending load was present, i.e., is there a specification that states the concentricity of the two surfaces?

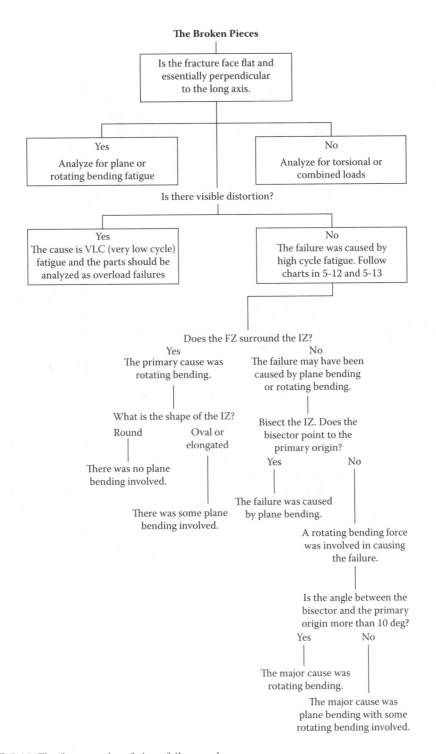

FIGURE 5.11 The first steps in a fatigue failure analyses.

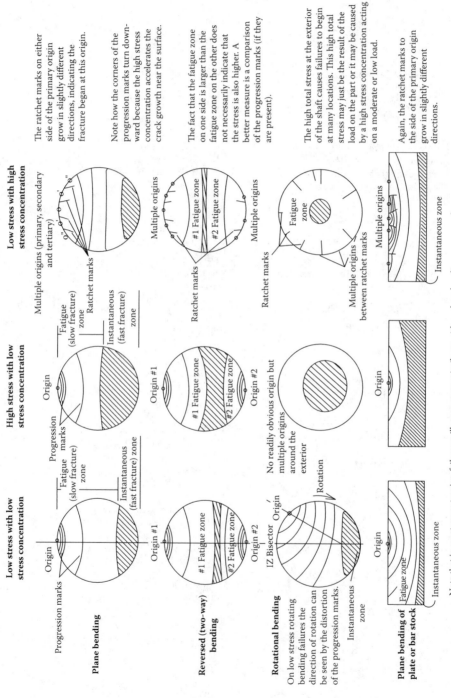

FIGURE 5.12 Fracture face appearances for rotating and plane bending failures.

Torsional failures in round stock

Torsional failure with relatively low stress and essentially no stress concentration

Torsional failure with relatively high stress and essentially no stress concentration

Triangular-shaped ratchet marks

Multiple origins

Torsional failure with low stress but high stress concentration

Torsional failures - in bar stock

Torsional failure with loading applied in one direction ranging from zero to a maximum. This example shows a moderate total load.

Torsional failure with loading applied in both directions, i.e. ranging negative maximum to a positive maximum. This example shows a very heavy total load.

Multiple causes

Combining rotating bending with plane bending or tension
Low stress and changing loads

Note how the bisector of the IZ doesn't point directly at the origin. This shows there is some rotating bending involved, but other forces have combined with it to cause the failure. The fact that the origin is so close to the bisector shows the plane bending was the greater of the two forces when the crack began.

Crack growth with reversing rotation

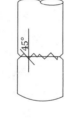

Notice how the progression marks grow more rapidly (wider spaced) on one side, then more rapidly on the other. This sketch was made from a shaft that had reversing loads and the irregular crack growth is the result of the changing direction of rotation.

Low stress, high stress concentration, and combined loads

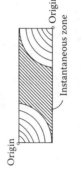

The oval-shaped instantaneous zone shows there is a combination of forces. (If there was only rotating bending the IZ would be round.)

The ratchet marks indicate the presence of multiple origins and high stress concentrations. The ratchet marks are approximately centered at the top and bottom of the shaft, showing the centers of their respective forces.

Combined torsion and either bending or tension

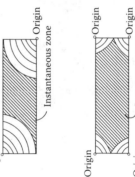

Combination of torsion and tension or torsion and bending forces. As the importance of the torsional loading becomes less, the fracture face more closely approaches the perpendicular.

FIGURE 5.13 Fracture faces from torsional and multiple load cases.

PHOTO 5.11 A rotating shaft that has failed. The keyway position, ratchet marks, and IZ shape all give clues to the failure cause.

6 Fatigue (Part 2): Torsional, Low-Cycle, and Very-Low-Cycle Failure Influences and Fatigue Interpretations

Chapter 5 gives the guidelines for the basic analysis of fatigue failures. Unfortunately, like much of life, there are many complicated differences between the "basic book" and what actually happens. In this section we'll build on those basics to give you the ability to solve actual failures in a better fashion.

TORSIONAL FATIGUE LOADS

In looking at those basic failures, most of the demonstrated loads involved plane- or rotating-bending forces. In the following paragraph we look at torsional fatigue forces and their interpretation.

Torsional fatigue loads tend to cause some confusion because our senses can't easily detect them. For example, we can easily see translational vibrations, such as a piece of sheet metal vibrating in response to some exciting force or a pump housing bouncing up and down because of cavitation forces, a misaligned coupling, or a weak base. However, we can't easily sense that the shaft in that pump is winding up and relaxing in response to a torsional exciting force. When we analyze the failure of a machine shaft, there has to be a torque applied to the shaft in order to have the shaft rotate, however, that doesn't mean there is a torsional fatigue load. The torsional fatigue load only comes into play when that torsional stress varies greatly in a relatively short time.

Diagnosis of torsional fatigue failures is similar to diagnosing any other fatigue failure. When the combination of fatigue stress and stress concentrators is low, there will be a single origin, and the fracture will grow on a 45° angle as seen in Photo 6.1.

This shows a 266 kw (200-hp) ductile iron compressor shaft that failed from pure torsional fatigue. There is a single origin with no stress concentration, and the fracture plane is at the classical 45°. But in this case, because the shaft is ductile iron and has poor crack toughness, the Instantaneous Zone (IZ) is both larger and rougher than what would normally be expected from a steel shaft.

If there had been a serious stress concentration on that shaft, the fracture would have developed in several locations almost simultaneously as shown in Photo 6.2. The cracks on this reducer input shaft developed on several of the splines and the failed piece looks almost like a star drill.

The process of diagnosing the causes of a torsional failure also has to include whether or not there is a bending force involved with that fracture and that involves looking at the plane of the failure as it progresses across the shaft as follows:

1. Single Origin
   a. If there is pure torsion involved with a single origin, the crack will grow at a 45° angle.
   b. If there are other forces, bending or tension involved, the angle will be less than 45°, roughly in proportion to the forces.

PHOTO 6.1 A classic torsional fatigue with a single origin. The coarse, rough IZ is because this shaft is ductile iron.

PHOTO 6.2 A torsional fatigue failure with multiple origins.

2. Multiple Origins
   a. If there is pure torsion, the cracks will grow toward the center, and the angle will be similar to that seen in Photo 6.2, or even steeper.
   b. If there are other contributory forces, the sides of the ratchet marks will be diagonals, as shown in Photo 6.3, and the fracture face will be close to flat.

PHOTO 6.3 This section of a failure shows many ratchet marks with angled sides indicating multiple torsional fatigue crack origins.

Photo 6.3 shows a reciprocating compressor shaft where the forces caused the operating failure of an improperly repaired shaft. The load on the shaft included the torsional driving force, torsional vibration from the irregularities of a reciprocating compressor, plus the weight of a huge motor rotor. The ratchet marks have diagonal sides, indicating that the cracking started from torsional loading, i.e., the driving force, in several locations. The main body of the fracture is relatively flat, showing that there was also a substantial plane bending load, i.e., the weight of the rotor, involved.

In the aforementioned examples, the failures were caused by torsional loads that were applied along the length of the shaft. An even more common location for torsional fatigue failures starts with the stress concentration at the root of a key seat. In a device that is driven by a key, the driving force is applied across the center of that key and, if there is a typically small radius at the bottom of the key seat, the resultant stresses can be substantial. Both Photo 6.4 and Photo 6.5 show the evidence of torsional fatigue cracking and in both cases poor coupling fits contributed to the failures.

Photo 6.4 is the input shaft to a small reducer. Photo 6.5 is the drive end of a large pump. In both cases, careless assembly practices contributed substantially to the failures. One of the problems with loose coupling fits is that the looseness results in fretting, which continually reduces the fatigue strength of the shaft material. Eventually, the strength drops to the point at which a fracture begins. (Many mechanics don't like to install couplings with interference fits because it entails more work, but the benefit is that the chance for fretting is eliminated, and the equipment lasts longer.)

PHOTO 6.4 The constant hammering from the loose coupling fit and the torsional fatigue forces both wore the side of the keyway and caused a crack that started at the keyway root.

PHOTO 6.5 The tapered-fit coupling wasn't properly installed and cracking started at the keyway root.

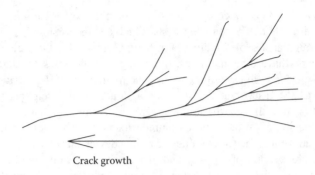

Crack growth

FIGURE 6.1 A sketch of a river mark.

RIVER MARKS AND FATIGUE CRACK GROWTH

The last of the fatigue failure surface features to be discussed are river marks. Figure 6.1 is a sketch of a river mark, and it looks like the symbol for a river on a map. There are many small tributaries to a typical river, and as the water flows downstream in the river, the crack growth also proceeds "downstream." They are actually created when cracks exist on several planes and gradually work together, but the value to the analyst is that they show the direction of crack propagation. They occur in all types of materials but are most frequently seen in the fracture of high-strength materials.

Photo 6.8 is an excellent example of river marks on a pump shaft failure face. These show the direction of the fracture growth changed dramatically in a short distance and from that we know the forces also changed.

PLATE AND RECTANGULAR FAILURES

There is a common lament from those that don't understand the basics of failure analysis; they want a catalog of "how to diagnose xxx" with a photo of the exact failure. What they don't understand is that most failure analyses aren't complicated, and diagnosing a plate failure, such as that in Photo 6.6, is essentially the same as the diagnosis of any other fatigue cracking.

This is a 50 mm (2-in.)-thick piece of stainless steel is from a large food product plant, and looking at it we see:

1. The progression marks and increasing surface roughness show that it is a fatigue failure.
2. The force that caused the failure is either tension or plane bending and, because the progression marks grow symmetrically for the first half inch, it was well centered on the origin.
3. The progression marks show there were a great many load changes during the crack growth.
4. The river marks show the fracture grew from the origin in the center generally toward the lower right showing that, as the crack grew larger, the force tended to be a little off center.
5. The small IZ in the lower-right corner shows the part was not heavily loaded at the time of final fracture.
6. An inspection of the surface at the origin shows the crack started at the base of a fretted pit (emphasizing the message in the preceding paragraphs about the problems that fretting causes).

Examining the knife blade in Photo 6.7 we see a fatigue crack with three general origins. The relatively uniform growth and the crack plane again state that the load was either tension or bending.

PHOTO 6.6 Plane bending failure of a rectangular section. This cracking originated in a fretted area of the top surface.

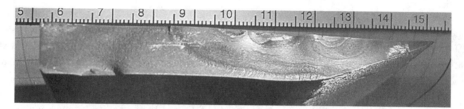

PHOTO 6.7 Plane bending failure of a hardened knife blade that was poorly supported.

PHOTO 6.8 The river marks show the crack took a sudden left turn.

(Usually determining which of these was responsible is a matter of a simple visual inspection. For example, it would be very difficult to get a tension load on a knife blade.) But two differences between this and the previous failure are:

- The off-center location of the origins indicates that either the stress was not uniform across the blade or there were some stress concentrations toward the tip.
- The large IZ indicates the blade was highly stressed.

FATIGUE DATA RELIABILITY AND CORROSION EFFECT ON FATIGUE STRENGTH

In Chapter 3, Figure 3.8 is an S–N curve that shows the fatigue strength of a metal when stressed at a given rate for a number of cycles. There is a tendency to believe that the material will always behave exactly as the curve shows; however, Figure 6.2 gives another more realistic view of an S–N curve. From this figure we can see that even though a piece may be loaded 10% less than the

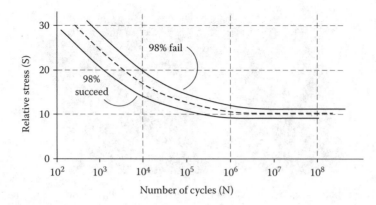

FIGURE 6.2 Confidence bands showing that the actual fatigue strength of a piece may vary by an order of magnitude.

average fatigue stress, it may still fail. Meanwhile, another seemingly identical shaft that is stressed to 10% more than the average fatigue stress doesn't break. The benefit of this chart is that it shows the variability in fatigue data and there are many applications in which the range, 1% survival and 99% survival, is an order of magnitude. Further complicating this situation is the fact that the fatigue strength can be greatly affected by the operating conditions.

For example, we know that corrosion decreases the fatigue life of a shaft, and we know generally that:

- The greater the fatigue strength of the material, the more rapidly it is reduced.
- The longer the part is attacked by the corrodent, the greater the decrease in strength and the greater the variability in strength.
- The more severe the corrodent, the more the strength is decreased.

But when we think about an 1800-r/min machine rotating 946,000,000 r/year, we can understand why there isn't a lot of data on the specific effects of various corrodents.

> Our general approach to solving corrosion fatigue problems is to try to eliminate the corrodent. Many times that is impossible and we turn to coatings or tapes to try to prevent corrosion from occurring, and we have had excellent results from using both anticorrosion tapes and high-quality electrical tape. We tend to keep away from changing materials unless we have absolutely positive data as to the full range of operating and shutdown environments.

The magnitude of this decrease is difficult to understand until you look at data such as that shown in Figure 6.3. This shows the fatigue strength of a 1.7 GPa (250-ksi) maraging steel that has been subjected to seawater at 25°C. One would expect the fatigue strength to be on the order of 615 MPa (90 ksi) in clean dry air, yet the data shows it isn't even 10% of that after less than 10,000,000 cycles — and that's not even a 10-d run on a 700-r/min machine.

Additional data can be seen in Chart 3.2 and most of the data we've seen for mild steels has indicated that, after 10^8 cycles in even mildly corrosive water with some chlorides, the fatigue strength has been cut at least in half.

RESIDUAL STRESS

A shaft or a beam is essentially unaffected by the source of the stress operating on it as long as the actual operating stress is well below the yield strength for reliable operation. The loading could

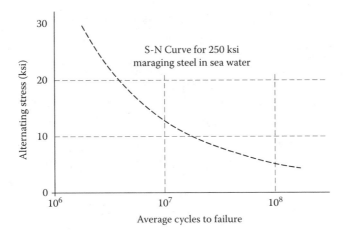

FIGURE 6.3 Notice the tremendous decrease in fatigue strength in this high-strength alloy.

be from many sources such as mechanical drive stresses, thermal stresses from temperature differentials, or residual stress from fabrication or treating processes.

There are times when residual stresses are beneficial, such as the compressive stress that results from thread rolling on a fastener. The compressive stress tends to offset the tensile stresses that cause fatigue failures, and the compressive stresses greatly increase fastener fatigue strength. Unfortunately, most of the residual stresses we see in plant machinery act to increase the tensile stresses, and they frequently contribute to fatigue failures.

For plant machinery, the most common sources of residual stresses all involve heating and cooling such as casting, welding, and heat treating. A good example of how residual stress can contribute to a failure is shown in Photo 6.9. One of the rules of analyzing a failure is that the crack always grows perpendicular to the plane of the maximum tensile stress. This figure shows a rocker arm that has a pair of fatigue cracks in it, and during normal operation the cracked area of the rocker arm should never see any tensile stresses, i.e., the push rod pushes upward on the right side and the valve spring pushes upward on the left side. The lower portion of the rocker arm and the area around the mounting stud will see tension. However, the cracked area should only see compressive forces, yet it had a fatigue crack.

This was an unusual failure and that area of the rocker was left in tension after heat treating. Then the normal operating stress put it into compression and, when the pushrod retracted, it went back into tension. It experienced tension and compression alternately more than 6000 times per minute and eventually suffered a pair of fatigue cracks.

PHOTO 6.9 Note the crack in the area that should never see tension during normal operation.

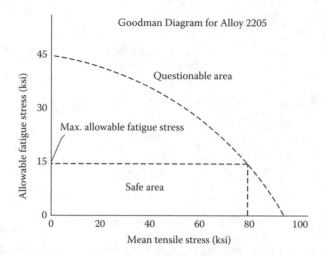

FIGURE 6.4 A Goodman diagram such as this allows us to calculate the combined effects of steady state and fatigue stresses.

COMBINED FATIGUE AND STEADY STATE STRESSES

There is a tremendous amount of data on the fatigue strength and the tensile strength of materials, but it isn't easy to understand how the two should be combined in an analysis. The first work on this was by Gerber in the late 1800s, using some of the data from Wohlber's early fatigue experiments. It generally remained a theoretical curiosity until aerospace applications in the 1940s and 1950s placed a premium on understanding exactly how they affect each other. It became a common aerospace technique in the 1960s but didn't reach the rest of the world until computers and FEA (Finite Element Analysis) arrived.

There have been other approaches to solving the combination of these forces and Figure 6.4 is called a modified *Goodman diagram*. (Goodman's approach improved on the earlier work of Gerber and Soderberg.)

The horizontal axis, the mean tensile stress, is a combination of all the steady state stresses operating on the part plus half of the absolute value of the fatigue stress. In practice the engineer would calculate these stresses and then draw a vertical line from that value to the Goodman line. Where the vertical line intersects the Goodman line, a horizontal line is drawn over to the vertical axis, the *fatigue (alternating) stress*. This is the maximum allowable safe fatigue stress on the assembly, and as the mean tensile stress is reduced, the amount of fatigue stress can be increased. In Figure 6.4 it can be seen that, if stress relieving reduced the mean tensile stress by 205 MPa (30 ksi), the allowable fatigue stress would more than double.

BASE MATERIAL PROBLEMS

It's not unusual to hear people blame failures on material defects (this certainly happens at times) but in reality it is not very common. Fewer than 1% of the failure analyses we have worked on have involved defects in the original shaft or billet material. Failures in castings and heat-treated pieces are a little more common, and weldment failures (involving human errors) are much more common.

When we started doing failure analyses in the 1970s, the percentage of base material failures was much higher but the quality of materials from all sources has improved steadily.

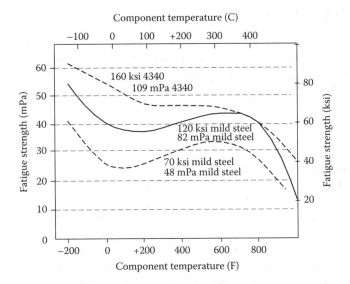

FIGURE 6.5 Fatigue strength vs. temperature for three steel alloys.

One of the areas where we see a great tendency to blame the material is when elevated temperatures are involved. The involved person will see that the part may be at 100° or 150°C (200° to 350°F) and think that the parts have been weakened, but, as can be seen in Figure 6.5, the actual fatigue and tensile strengths are either unaffected or slightly increased over this range.

VERY-LOW-CYCLE AND LOW-CYCLE FATIGUE

Most fatigue failures result from high cycle fatigue and involve millions of cycles between the initial force application and the final failure. However, looking at some other fatigue failures it is apparent that they did not take a very long time. For example, if you take a paper clip and bend it back and forth it will eventually break. Sometimes it only takes four or five bends and other times it takes 25 or 30. In both situations the paper clips fail from fatigue but it has taken a lot less than the millions of cycles discussed in Chapter 5 on high cycle fatigue.

Figure 5.1 shows the commonly accepted regions of Low Cycle and High Cycle Fatigue (HCF). As shown, there are differing opinions about whether the range between 10^4 and 10^5 cycles should be considered high or low cycle fatigue. In a similar manner, in the area below about 10 of 20 cycles there are differing opinions on the failure terminology. However, from a practical viewpoint, this is Very Low Cycle Fatigue (VLC) and, compared with HCF, there is a tremendous difference in fracture mechanisms.

With HCF, failures take a relatively long time to occur. Cracking begins after the atomic dislocations line up and create a plane of weakness and the crack slowly propagates across the piece leaving very little in the way of deformation. Researchers have found that, for an HCF failure, typically about 90% of time is spent on crack initiation, leaving only 10% for the actual fracture growth.

VLC failures occur at loads close to what would cause an overload failure, usually above the yield strength and frequently close to what would cause a tensile failure of the part. As a result, the relative ductility and/or brittleness of the material plays an important part in the failure's appearance. Most engineering materials are ductile and, with VLC fatigue, as the part is stressed there is deformation and the grains of the part are literally torn apart. The result is a fracture that takes relatively few cycles, almost always less than 20, with the reassembled pieces usually showing some slight deformation but much less than would be seen with a ductile overload failure.

VERY-LOW-CYCLE IN RELATIVELY BRITTLE MATERIALS

Photo 6.10 shows a portion of a broken fork from a medium-sized lift truck. On inspection, one can see that there is a small fatigue zone (FZ) all the way across the top of the fracture face with the remainder a very large IZ. A close inspection of this FZ shows at least seven origins, each with a series of four or five semicircular progression marks. Forks are designed with reasonable safety factors, and we can tell that from this that the failure was:

1. Not instantaneous with the first high-load application, i.e., it has been overloaded in the past.
2. Grew in at least five cycles before it snapped like a brittle fracture.
3. Was very heavily loaded at the time of final fracture.

Comparing this with the fracture face seen in Photo 4.17, we see that fork has five notchlike cracks that had existed for a long time when it was overloaded and broke. It may be a trivial point because they both broke after relatively few stress cycles; however, the notches in the fork in Photo 4.17 are absolutely uniform and showed no evidence of a growth pattern prior to the fracture, whereas the cracks in the piece in Photo 6.10 saw at least five overload cycles before fracture occurred.

A second example of a VLC fracture can be seen in the steel-mill-forming roll shown in Photo 6.11. The roll had been used for one short run, taken out of service for a week, and then put

PHOTO 6.10 A lift truck fork that failed from repeated overloads.

PHOTO 6.11 A roll from a steel mill. The blackened area is an older oxidized crack surface. The IZ is the lower third of the roll.

back in run shortly before it broke. Looking at the photo, the initial FZ is the blackened area centered at about 11:30 o'clock. and an inspection of the fracture found the following:

1. An extensive network of grinding cracks on the radius.
2. There was some major cracking with three obvious progression marks during the first run.
3. After the unit was put back in run, it was very heavily loaded, and the fracture propagated about 23 cm (9 in.) in one revolution.
4. That was followed by four progression marks about 2 cm (0.8 in.) apart, shortly after which the roll broke in two.

From the analysis we could tell the client that, yes, they did load the roll very heavily, but the primary reason it broke was the stress concentration resulting from the grinding cracks.

VLC IN DUCTILE MATERIALS

The fracture shown in Photo 6.12 looks suspiciously like a torsional ductile overload failure. It is from the input of a large machine, and close inspection of the face shows a substantial amount of smearing as the pieces twisted apart. The clue that it is VLC lies in looking closely at the OD of the fracture, the area from 11 to 1 o'clock and the area from about 6:30 to 7:30 o'clock, where several fatigue origins are evident. In order for these to occur, the unit had to have been heavily loaded. Our suspicion is that the failure probably took about eight to ten fatigue cycles from the first significant load to final fracture, but the questions that have to be addressed at the plant site are, "Did this tremendous fatigue overload occur immediately before failure or was it a recurring event? Did the failure occur from a series of eight or ten immediately sequential loads or were the loads applied at widely differing times?"

The shaft shown in Photo 6.13 took longer from initiation to final fracture and really belongs in the category of low cycle (LC) and not VLC, but this is a good place to introduce it. Examining this paper machine drive shaft we see the perimeter is ringed with ratchet marks showing numerous failure origins. Two thoughts to keep in mind during the diagnosis are:

PHOTO 6.12 The very low cycle torsional fatigue failure of a large machine shaft.

PHOTO 6.13 This low cycle torsional fatigue failure saw more stress cycles than the previous shaft.

1. Paper machine designs are relatively conservative.
2. When the part failed, the operations personnel were starting the machine section. The start-up procedure consisted of clutching the drive in and out until the machine got up to speed.

This shaft has a diameter of about 15 cm (6 in.). The multiple ratchet marks have diagonal sides indicating they were caused by torsional loads and were surrounded by relatively smooth areas showing that they had existed for some time. Inspection shows the crack then propagated inward for about 1 cm (0.4 in.) when the loading changed rapidly and the central 10 cm (4 in.) was the IZ.

The diagnosis was that the shaft was turned off because of the start-up procedure that aggravated torsional resonances. The failure analysis indicated that this was probably the second such event and that the plant should revise their operating procedures. (They had an identical failure 2 years later.)

UNUSUAL SITUATIONS

In the chapter on overload failures, there were several examples in which ductile materials behaved at least partially in a brittle manner because of the rate at which they were loaded. Similarly, there are times when VLC fatigue failures have unusual appearances.

One of the most interesting and unusual situation occurs when there are alternating bands of ductile rupture and cleavage on the fracture face. Photo 6.14 shows the fracture of a 10-cm (4-in.) low-speed drive shaft for a curing oven that was having a problem with the oven chain snagging, i.e., periodically seizing in position and then breaking loose. Looking at the fracture, in the upper-left corner of the keyway there are a series of fatigue striations over about 5 mm (1/4 in.) followed by a 2-cm-wide (0.8-in.) brittle fracture area. This was followed by a 1-cm-wide (0.4″) area of progression marks and then the entire shaft snapped off. Interpreting these, we see:

1. There was a heavy load that caused the initial 5 mm of fatigue cracking. Also, it was aggravated by the stress concentration at the end of the keyway and the corrosion on the exterior of the shaft.
2. The chain snagged, and this greatly increased the load on the shaft, but the chain broke loose before there was enough stored energy in the shaft to propagate the crack all the way across.

PHOTO 6.14 An oven chain drive shaft with an unusual surface appearance.

3. The extremely rapid crack growth stopped after 2 cm (0.8 in.) and then slowly continued to propagate by fatigue, developing the progression marks.
4. The chain snagged again and this time the shaft broke.

FAILURE EXAMPLES

To understand how a failure occurs, start with the diagram in Figure 2.5 to understand the basic type of failure. If it is diagnosed as a fatigue failure, then look at Figure 5.11 to understand the forces involved and interpret the surface features.

EXAMPLE 6.15

This is a 200-kW (150 HP), 1800-r/min motor shaft from a paper mill. It had been in run for about 6 weeks after a repair when the shaft broke Photo 6.15.
    Inspection — from the failed piece and the significant points on the sketch:

A. The fractures began on both sides of the key. We can tell this because there are ratchet marks on both sides of the key and also because this is the smoothest part of the fracture face. There is some curvature along the edge indicating a high stress concentration at the origin.
B. This is a progression mark. It is symmetrical indicating plane bending as the major fracture cause.
C. Another symmetrical progression mark indicating plane bending. The area between the two progression marks becomes rougher as the crack progresses from B to C. The fact that it is not very smooth indicates a fast-growing crack.
D. The IZ is relatively small, indicating a light load, and oblong, indicating a combination of plane and rotating bending.

Diagnosis and Recommendation

With a motor shaft used for a belt drive we would expect a rotating-bending load. However, after the cracking started, both the progression marks are symmetrical, and this indicates that the major stress was plane bending. As the failure progressed, the shaft became weaker and the rotating-bending component of the stress increased. At the final fracture, the IZ bisector did not point at the origin, indicating that there was a substantial rotating-bending component. But the elongated shape of the IZ indicates most of the loading was from plane bending. Therefore, our opinion is

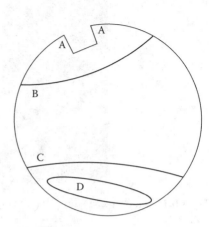

PHOTO 6.15   **FIGURE 6.15**

that the major physical cause was that the shaft was bent and a secondary cause was the rotational loading.

The recommendations to the motor shop were:

- Improve (or start) a procedure for checking shaft run-out before the motor is shipped.
- Increase the radius at the step in the shaft, reducing the stress concentration.

For the plant, we suggest checking and recording the sheave speeds and the tension ratio on the belt drive.

EXAMPLE 6.16

As shown in Photo 6.16, This is a paper mill refiner shaft and the major load should be torsional with very little rotational bending. It operates at 885 r/min and is approximately 5 years old. Major machine maintenance had taken place about 2 months earlier.

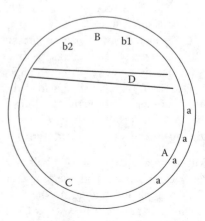

PHOTO 6.16   **FIGURE 6.16**

Inspection

These fatigue cracks began at two places A and B. However, start by looking at the step in the shaft, and we see there is obvious corrosion pitting, shown as "a" in four locations reducing the fatigue strength of the steel. Then we see a ratchet mark at A, indicating that there were actually two crack origins, one on each side of the ratchet mark. Looking at the progression marks surrounding these origins, we see that they are not symmetrical, which indicates some rotating bending is involved. Next, at B there are also two origins centered on a ratchet mark one third of the way from b1 to b2. The ratchet marks at these two positions are pointing away from each other, proving that there is a fracture origin between them. Furthermore, there are several faint asymmetrical progression marks between these origins and the IZ at D. The rough area at C is uncommon, but the river marks indicate that it was one of the last areas to fracture before the shaft came apart at D. D is the final fracture and is relatively rectangular, indicating the final fracture was from plane bending.

Conclusions and Recommendations

Interpreting this complicated but relatively straightforward failure, we can see that corrosion greatly reduced the fatigue strength of the shaft. Then the cracking started, first at A and then at B. Both of the fracture planes propagated primarily by rotating bending but with a significant component of plane bending until the final fracture when the shaft was not heavily loaded.

Our recommendations to the people at the plant were:

- Improve the shaft alignment to eliminate the rotating-bending load.
- Tape or coat the shaft to try to reduce the fatigue strength deterioration from corrosion.
- Add a concentricity specification to the shaft drawing to reduce the plane-bending load.

EXAMPLE 6.17

This was the stem for a 2-m-wide (6-ft) gate valve in a sewage treatment plant. It failed 3 months after installation, and we were asked to diagnose what had happened. The stem was driven by an electric motor through a reducer and rotated to lift and lower the gate. The gate was positioned vertically and designed to move up and down in a narrow gap between guides.

PHOTO 6.17

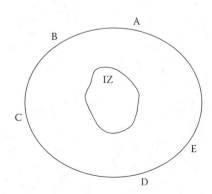

FIGURE 6.17

Inspection

The examination began with noting that the surface of the machine's stainless steel shaft was very rough. Then, examining the fracture face found that there are five general areas of origins at A through E. Along with those origins were some prominent ratchet marks, showing the fractures began on different planes. Next, looking at the progression marks, it was seen that they grew inward symmetrically from each origin. The final fracture zone is about what would be expected in a piece this size.

Examination showed the stem was being flexed back and forth while in position. The vibration from the flexing allowed the reducer to slowly allow the gate to drop, repositioning the stem and creating another crack.

Conclusions and Recommendations

- The gate guides were not properly installed and allowed the gate to move, flexing and heavily stressing the stem.
- The rough machined surface resulted in small stress concentrations that hastened the failure.

EXAMPLE 6.18

The shaft as shown in Photos 6.18 and 6.19 has a diameter of about 12 cm (4¾ in.) and rotated at almost 600 r/min. It was used in a process plant and is an interesting failure because of the shape. The sketch shows a cross section of the piece, and the operating details are not as important as understanding the surface features.

Inspection

Inspection shows cracking began around the entire shaft at the shoulder radius. The multiple ratchet marks point toward a high stress concentration, and the steeply angled fracture face is perpendicular to the stress field. The early fatigue zone is smooth around the entire shaft. It then grows into a

PHOTO 6.19

PHOTO 6.18

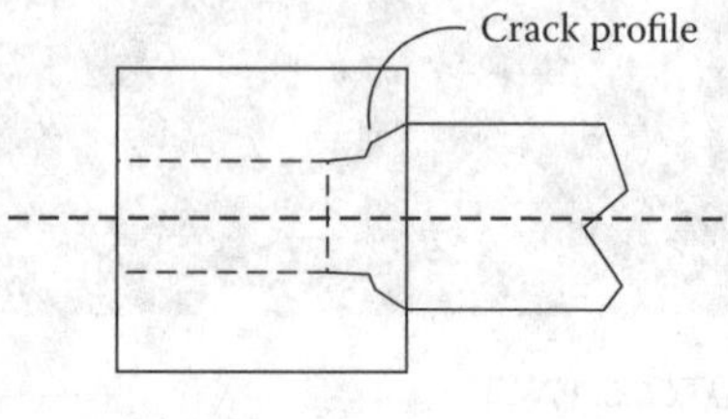

FIGURE 6.18

Crack profile

much rougher and faster fracture with numerous river marks and then into an IZ that is parallel to the shaft axis.

From the fatigue zone it appears that the only stress on the shaft was rotating bending, and the multiple origins around the shaft show the product of stress and stress concentration is very high. There are numerous progression marks and this indicates the load is changing, but the size of the IZ indicates it was very heavily loaded at the final fracture.

Conclusions and Recommendations

- Improve the alignment to reduce the rotating-bending stress on the shaft.
- Shorten the inner bore, allowing much more room between the stress concentrations from the internal and external steps in the shaft.
- On both steps, make the transition radii larger to reduce the stress concentrations.

The rough, almost woody appearance of the IZ lead many people to suspect that the material was defective. However, this is the typical appearance of an IZ or a brittle fracture that propagates in the same direction that the piece was rolled. We are used to seeing IZs that run perpendicular to the rolling direction and they produce rough crystalline appearing surfaces. When they grow in the rolling direction, they tend to fracture along the grain flow lines and the result is something that looks like a piece of petrified wood.

EXAMPLE 6.20

Both Photo 6.20 and Photo 6.21 show views of an 8.6-cm (3.5-in.) drive shaft from a process winder. The plant had had problems with short coupling life and had recently replaced the coupling with a larger one. The coupling fit is at the far-right side, the last cm (0.4 in.) of Photo 6.20, with the winder toward the left side.

Inspection

The diagonal fracture face shows this is a torsional failure. Then, examining the surface we see that:

1. The surface gets progressively rougher as it travels across the shaft, but even at the beginning it is not very smooth.
2. As shown in Photo 6.21, the cracking starts at the upper corner of the end milled keyway.
3. There are numerous progression marks indicating many load changes.
4. The final fracture zone, the IZ, is relatively large.
5. There is obvious fretting at the end of the key fit.
6. The coupling fit appears to be tight with only very slight fretting on the drive side of the interference fit.
7. Hardness testing showed the tensile strength of the shaft was about 80,000 psi.

PHOTO 6.20

PHOTO 6.21

Conclusions and Recommendations

This is a pure torsional fatigue failure with a very high torsional load. The coupling fit was excellent and did not contribute to the failure. The fact that the previous couplings had also been destroyed leads us to suspect there is a torsional resonance in the shaft.

We recommend further analysis to understand the shaft loads, i.e., checking to see if there is a torsional resonance and also taking high-speed-recording ammeter readings on the motor to see the magnitude of the motor loads. With that combination, they should be able to see how highly the shaft is loaded and how the loads vary. Possible corrective actions could include changing the torsional stiffness of the assembly by changing the type of couplings and strengthening the shaft, but an understanding of the true forces on the shaft is a necessity.

7 Understanding and Recognizing Corrosion

Of all the failure mechanisms, corrosion is probably the most expensive. For the past 60 years, studies by the Battelle Institute and other organizations have shown the U.S. spends somewhere between 3 and 4% of their GNP on corrosion prevention, treatment, and correction. For most of the developed world the cost is similar; however, the warmer and more humid countries are even more burdened.

The reason corrosion exists is shown in Figure 7.1. We take minerals out of the ground and expend a tremendous amount of energy to make them into useful metals. This increases their potential energy and Mother Nature tries to reclaim that energy over the metal's life. It is interesting to note that the metals found naturally occurring are generally those that have a great deal of corrosion resistance, such as silver and gold. The rest of the metals we use are mined as oxides.

The key to understanding corrosion is to realize that it is the result of electrical currents and electrochemical reactions. Different types of corrosion cause those currents to be initiated in slightly different ways, but a thorough understanding of the current generation and conduction (and the reaction kinetics at the anode and cathode) is a necessity to understand and measure corrosion.

Most corrosion requires that some moisture be present but dry corrosion also exists. It is not very common and almost always happens at elevated temperatures. A simple example can be seen by heating a mild-steel rod with a torch until it is red hot, and when it immediately develops a flaky gray-black oxide scale. Technically, this oxide scale is corrosion, and the name for this dry corrosion mechanism is high-temperature direct conversion.

With the exception of these high-temperature direct-conversion examples, and possibly some of the hydrogen damage mechanisms, one of the more important points to remember when trying to solve corrosion problems is that there must be moisture present in order for the attack to occur.

Figure 7.2 is an example of a typical corrosion reaction. It shows a zinc bar, but it could be any metal in a conductive electrolyte. For the reaction to take place, there has to be an anode and a cathode, and a liquid is needed to conduct the corrosion currents from one to the other. The anode is attacked and releases metal ions into the liquid to form oxides.

Also, as shown in the figure, at the cathode the corrosion reaction produces atomic hydrogen, H^+. In most cases, these hydrogen atoms immediately unite with other hydrogen atoms to create the common molecular form, hydrogen gas, H_2. But the fact that hydrogen is generated in the atomic form is the source of many of the hydrogen damage problems. In understanding the mechanisms of hydrogen damage, it is helpful to realize that essentially steel is to hydrogen as chicken wire is to feathers. The chicken wire will slow down the feathers but it won't stop them from passing through. In a similar manner, the hydrogen atom is so small it can wander through the interstices in the steel. When it finds and joins with another hydrogen atom, the molecular forces result in hydrogen gas, and substantial stresses are created.

A second point to realize about this hydrogen generation is that there will always be some dissociation. That is, although most of the hydrogen atoms unite to form hydrogen gas, some of them will remain as ions. It is these ions that result in the damage. A third important point is that there are chemicals such as sulfur that tend to aid in this dissociation and act to restrict the formation of hydrogen gas.

The idea of free hydrogen atoms and that there are always some present can be a difficult one to understand. One interesting example that will illustrate this is the degradation of titanium heat

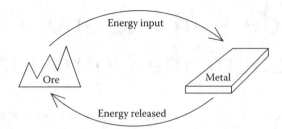

FIGURE 7.1 The basic cycle with corrosion returning the metal to an oxide.

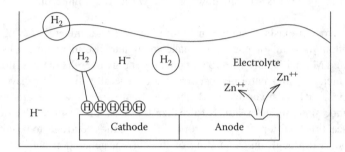

FIGURE 7.2 An idealized view of a zinc bar corroding in a liquid. At the cathode hydrogen atoms are released, while the anode is attacked and releases metallic ions.

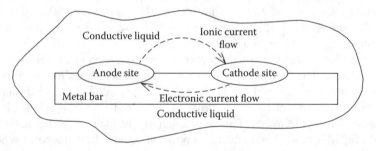

FIGURE 7.3 A diagram of the current flow in a corrosion reaction.

exchanger tubes in a plant we worked with. The vessels were large shell-and-tube heat exchangers with steam on the shell side and process liquor on the tube side. The tubes were welded titanium, and they would fail after about 7 years. The failure mechanism was true hydrogen embrittlement, i.e., the free hydrogen atoms from the steam would unite with the titanium and slowly form titanium hydride, a brittle ceramic-like material. Eventually, the entire tube wall would be converted to this brittle material, and the tube would crack and leak. But the portion of the tube outside the tube sheets, in the water box, was completely unaffected and remained as ductile as a paper clip! The culprits were the free hydrogen atoms in the steam.

Looking at the reaction in Figure 7.2, a conductive liquid has to be present to complete the circuit. The corrosion current travels through the anode and cathode and returns through the liquid. If the liquid isn't present, the circuit isn't completed, and the reaction stops. Taking a closer look at that bar, Figure 7.3 shows how the currents travel.

Some important comments about the liquid are:

1. If the conductive liquid were prevented from getting to the surface, there would be no corrosion — because there would be no way to conduct the corrosion current. (This is the basis of many corrosion prevention approaches.)

2. In a similar manner, if the liquid contact with the surface were slowed, the reaction rate would also become slower.

3. If the liquid contained little or no oxygen, the corrosion rate would be greatly slowed.

4. The liquid doesn't have to be water and doesn't have to be at low temperatures. An example of this occurs in coal-fired boilers where pitting corrosion occurs when the sulfur compounds in the coal liquefy and carry the corrosion currents at temperatures well above (1000°F) 550°C.

5. Atmospheric corrosion tends to start in nooks and crannies and at surface irregularities. Research has shown that first traces of corrosion generally begin at about 60% relative humidity and get worse as the humidity increases.

CORROSION RATES

Many conditions affect corrosion rates including temperature, exposure time, moisture and liquid conductivity, pH, oxygen availability, solution contaminants, flow velocity, surface conditions, etc. The reactions are extremely complex and dependent on multiple variables. Because of the number of variables and their interactions, it is difficult to use laboratory experiments to accurately reproduce the actual field corrosion rates.

pH EFFECTS

Figure 7.4 shows a pH scale. The pH of a chemical is a measure of the relative number of hydrogen ions, and it is the common measurement used to express whether a material is acidic or alkaline. The range of the scale is from less than 1 up to 14, and a low pH indicates the material is very acidic. Conversely, a high pH indicates that the material is alkaline (or basic).

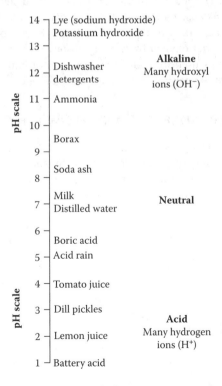

FIGURE 7.4 A pH scale with some common chemicals.

The pH of a corrodent can have a tremendous effect on the corrosion rate of a metal. Normally the lower the pH of a material, the more corrosive it is, and, usually, acidic materials are much more corrosive than neutral materials. In general terms, the acid tends to destroy the oxide coating on a metal and increases the amount of corrodent that can reach the surface.

Although corrosion rates drop as the pH approaches 7, in many cases they begin climbing again when the solution becomes highly alkaline. However, neither highly acid nor highly basic solutions are universally corrosive. For example, nickel alloys are highly resistant to very alkaline solutions, whereas the same materials rapidly corrode aluminum alloys. Diagnosing the specific effect of a pH requires research into the detailed conditions.

EFFECT OF AVAILABLE OXYGEN

Figure 7.5 shows a general chart of the effect of oxygen concentration on corrosion rates. Oxygen has to be present for oxidation (corrosion) to occur, and the greater the oxygen availability, the greater the corrosion rate.

Some practical examples of the effects of oxygen concentration and availability on corrosion are:

- In many tanks and storage vessels, the area of greatest corrosion is in the range of the normal operating liquid level — because this is where the moist surface can readily absorb oxygen from the atmosphere.
- In marine applications, the area of greatest corrosion is in the splash zone, with readily available oxygen, whereas areas deep underwater where there is little oxygen tend to see much less corrosion.
- An unusual example of the importance of restricting the availability of oxygen to both the corrodent and the surface was in a series of large mild-steel liquid calcium chloride storage tanks. The tanks used a layer of fuel oil floating on the surface to not only coat the sidewalls of the tank (and restrict oxygen from getting to them as the liquid level varied) but also to restrict oxygen from being absorbed by the solution.
- One of the mechanisms used to reduce the corrosion of boiler water is to deaerate the water and remove the oxygen.

EXPOSURE TIME AND FLOW EFFECTS

In general, corrosion rates tend to decrease with time. Figure 7.6 is an idealized diagram but closely resembles the typical decreasing corrosion rate. With many solutions, a buildup of hydrogen gas at the cathode or a buildup of corrosion products at the anode forces the corrosion rate to slowly decrease. But if local conditions affect these buildups, the corrosion rate can vary tremendously.

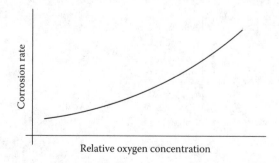

FIGURE 7.5 As oxygen becomes more readily available, most corrosion rates increase.

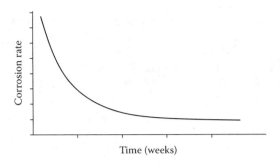

FIGURE 7.6 Over several weeks corrosion rates usually stabilize and then remain constant, providing the conditions don't change.

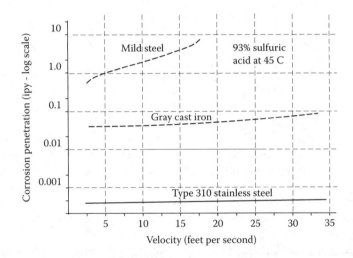

FIGURE 7.7 Corrosion rates vs. solution velocity for three materials with differing corrosion mechanisms.

In understanding how those local conditions can change the corrosion rates, we frequently see substantial variations as a result of the corrodent flow rate and some of these can be seen in Figure 7.7. This shows the effect of flow rates on the corrosion caused by 93% sulfuric acid. The three materials in the example corrode at greatly differing rates, and the interesting points are both the differences in the basic corrosion rates and how those rates vary with velocity. The mild-steel attack varies tremendously as the velocity changes, but both the gray cast iron and the 310 SS are almost unaffected. The reason for this is that the three materials develop their corrosion resistance by different mechanisms.

Carbon steel is frequently used as a 93% sulfuric acid storage material, and at zero velocity and 45°C (113°F) the corrosion rate in 93% H_2SO_4 is less than 0.25 mm/yr (0.010 in.) per year. But at 15 ft/min it is corroding at almost an inch per year — because the flow velocity scrubs off the protective ferrous sulfate layer, constantly exposing a bare steel surface to fresh corrodent, and the corrosion rate never drops off. Meanwhile, the gray iron is corroding at less than 1/10th the rate of the mild steel, because the free graphite in it tends to protect the surface and slows the acid attack. And the 310 SS, which relies on a durable chrome oxide coating to resist the acid, is almost unattacked because the acid can't penetrate that tenacious oxide layer.

At room temperatures, 93% sulfuric acid steel initially corrodes very rapidly. It forms a tight layer of ferrous sulfate, which reduces the amount of acid that can get to the steel, and the corrosion rate then drops rapidly to a very low rate. But if this protective layer is disturbed, it exposes the steel and the corrosion continues. In inspecting sulfuric acid storages, the important places to look are:

1. Those areas where there could be flow disturbances such as a fill pipe near the tank wall or an outlet pipe where there is an elevated flow velocity.
2. The areas where the hydrogen bubbles, from expected slow corrosion, could flow together and scrub away the protective sulfate layer. This damage is frequently seen in the heat affected zones of vertical welds and along the top of long horizontal steel pipe runs.

TEMPERATURE EFFECTS

Other than the presence of moisture, we can think of nothing that affects corrosion rates as much as temperature.

A good place to start in understanding these effects is Figure 7.8. This is a chart of corrosion rate of a metal in water vs. temperature but doesn't show values for the temperatures. Looking at it we can see that as the temperature drops to point A, the corrosion stops. From this we know that point A is 0°C, below which there is only ice and ice can't conduct the corrosion currents. Then, as the temperature increases and reaches 100°C, the liquid turns to gas at point B and the corrosion stops again — this time because gases can't conduct the corrosion currents. An ironic point about this chart is that if we add salt to the water, such as deicing road salts, it adds to the corrosion not only by decreasing the freezing point, but also by adding contaminants to the water.

Looking further into the effects of temperature we should touch on at the work of Svante Arrhenius, a Swedish Nobel Prize winner who spent a goodly part of his life publicizing the fact that, for most chemical reactions, in the range of temperatures around the ambient temperature, for a 10°C (18°F) increase in surrounding temperature, the rate of a reaction approximately doubles.

Arrhenius' equation essentially states that the rate constant, a key factor in calculating the rate at which a reaction will proceed, is an exponential function of the absolute temperature. Understanding it is helpful when it comes to corrosion rates, but it is invaluable for understanding any of the degradation mechanisms that involve chemical reactions. We refrigerate food to slow the reaction rates. We wash clothes in hot water, not because the detergents can't clean them in cold water, but because the detergents act so much more rapidly in hot water. As temperatures go up, electronic

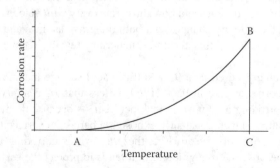

FIGURE 7.8 Corrosion rate vs. temperature.

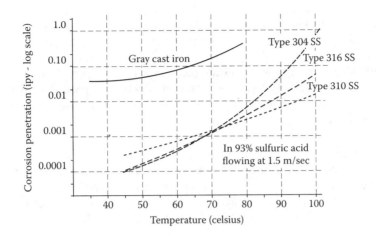

FIGURE 7.9 Corrosion rates vs. temperature for two of the materials shown in Figure 7.7.

components die more rapidly, polymer products such as seals and belts degrade more rapidly, and corrosion rates almost universally increase, approximately doubling with every 10°C increase.

In Figure 7.9, the effect of temperature on the corrosion of various alloys is shown. Again, two valuable points we can glean from this graph are the differences in corrosion rates with temperatures and the effects of the different alloying elements. The gray cast iron, even with the relatively high acid flow rate of 1.5 m/s (5 ft/sec) does relatively well at 45°C, but when the temperature increases by 30°C (54°F), the rate of attack has more than doubled. Meanwhile, the attack on the Type 304L SS goes from 0.002 mm to almost 2.5 mm (0.0001 in./year to almost 1 in./year) — more than Arrhenius would have predicted, but the Type 310 follows the rule almost exactly.

THE EFFECT OF CONTAMINANTS

In understanding how the complexity of corrosion is multiplied by contamination, it is helpful to look at atmospheric corrosion. There are available atmospheric corrosion maps of most of the industrial world and three broad area categories and their corrosion rates are:

- Rural
  - Subarctic — 0.0025 to 0.025 mm/yr (0.1 to 1 mil/year (mils per year, where 1 mil = 0.001 in.))
  - Temperate — 0.008 to 0.08 mm/yr (0.3 to 3 mils/yr)
- Industrial — 0.18 to 0.38 mm/yr (7 to 15 mils/yr)
- Marine — 0.10 to 0.60 mm/yr (4 to 24 mils/yr)

Looking at the data for rural areas, it is fairly obvious that the subarctic areas are drier and colder than the temperate areas and the rates are correspondingly lower. However, the corrosion rates for industrial areas, even though most are in temperate zones, are increased because of the industrial discharges, the sulfates, chlorides, and other chemicals. In marine areas, where there is plenty of moisture available and frequent contamination from a high salt content, the corrosion rate is even greater.

The two specific contaminants that cause the most problems are sulfides and chlorides. An example of the effect of chlorides can be seen by looking at the corrosion rate of Type 304 stainless steel in sulfuric acid. Normally the chrome oxide coating that develops on the stainless is adequate for protection against the acid attack. But if even a small amount of hydrochloric acid is present, the oxide film is attacked and the corrosion rate is increased tenfold or more.

> In one job we worked on, the personnel specifying the materials of construction on a 5000-hp centrifugal compressor didn't notice that the process gas was contaminated with about 0.2% hydrogen sulfide. They specified a stainless steel for the rotor material and after several months of operation and repeated vibration problems, found the corrosion rate was 3.2 mm/yr (125 mils/yr). The multimillion-dollar rotor was only 6.5 mm (¼") thick to start out with. A change in rotor material reduced the corrosion rate to less than .3 mm/yr (10 mils/yr), but it was an expensive lesson.

An example of the confusion and variables involved with material concentrations can be seen by looking at the effect of concentration on the corrosivity of sulfuric acid in carbon steel. As mentioned above, concentrated 93% sulfuric acid is frequently stored in carbon steel tanks and the tanks can be expected to last for more than 25 years as long as the flow velocity is low. Yet less concentrated sulfuric acid, for example a 50% concentration, would destroy the same tank in a few weeks.

Another interesting example how solution concentrations can play an important role in corrosion rates is the *hot wall effect*, a situation where evaporation concentrates the remaining solution. For example, if there was a liter of an aqueous solution with 1% of a corrodent in the solution and evaporation reduced the total volume to one-half liter, the corrodent concentration would be doubled. As the evaporation continues, the solution becomes proportionally more and more aggressive. The action of this hot wall effect is commonly seen in heat exchangers, but two examples in which out-of-service equipment failed as a result of it are:

1. A large assembly of Type 304 SS pipes destined for a turbine oil system was hydro tested using domestic water, then allowed to sit in a field for several months waiting for installation. After the test, the water wasn't completely drained, and the resulting evaporation allowed the solution to become concentrated to the point at which it enabled the residual fabrication stresses to cause stress corrosion cracking.
2. A heat exchanger was taken out of service and cleaned, but the tubes were not blown dry. The normal tube sag trapped some solution, which, together with the evaporation, resulted in internal pitting of the tubes.

EIGHT TYPES OF CORROSION — RECOGNITION AND PREVENTION

Depending on who you listen to, there are somewhere between eight and eighty forms of corrosion, but what we have tried to do is use some of the general groupings and make it understandable for the person who is not a corrosion specialist. We prefer the classifications proposed by Dr. Mars Fontana, and list his eight major types of corrosion and some information that should help you identify why the attack is occurring and what can be done to help prevent it. These corrosion mechanisms can be divided into two categories as shown in Figure 7.10.

Of these mechanisms only uniform corrosion results in overall attack of the piece. All of the others are relatively localized.

A point to keep in mind is that corrosion mechanisms frequently work synergistically, i.e., pitting can help cause stress corrosion cracking, etc. This frequently adds confusion to diagnosing and solving the causes.

UNIFORM CORROSION

This is the most common type of corrosion and probably causes 80% of the cost of all corrosion. One of the confusing points about it is that the name "uniform corrosion" refers to the mechanism and not the appearance.

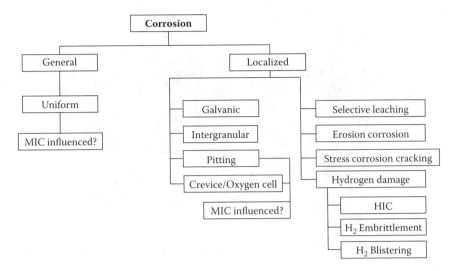

FIGURE 7.10 A breakdown of the major corrosion categories.

Appearance

This is the rusting (or oxidation) frequently seen on an exposed area of steel. The attack may appear relatively uniform, but more often there will be irregularities caused by local conditions and the surface won't be smooth. Photo 7.1 shows a pair of slip-joint pliers with corroded bands on the face. Referring back to Figure 7.2, one can see the anode areas are rusted, whereas the cathode areas are still clean and like new. Yet we know that 6 months from now the pliers will be rusted all over. This is a great example of uniform corrosion where today's anode is tomorrow's cathode, and the two poles constantly shift back and forth as the entire piece slowly rusts away.

This is the major mechanism that causes rusting of cars, bridges, equipment left outside, etc. Photo 7.2 is an example of uniform corrosion that (typically) doesn't look very uniform and shows a railroad bridge support in a large Midwestern city. The corrosion damage at the base of the column is the result of the city failing to both paint the column and to clean debris collected around it for a long time. The debris retains the natural condensation and is loaded with contaminants that help maintain the corrosive attack.

Of the major forms of corrosion, the penetration rate of uniform corrosion is the easiest to determine and predict. Because of this relatively predictable rate of attack, regular inspections should be able to avoid unexpected failures.

PHOTO 7.1 Uniform corrosion of a forged piece, showing anode and cathode areas.

PHOTO 7.2 Moisture and contaminents present in the debris around this bridge support lead to the attack.

PHOTO 7.3 The increased velocity under the baffle plate on this heat exchanger caused increased corrosion rates.

But uniform corrosion is also dependent on flow velocity. Photo 7.3 shows a large heat exchanger with severe erosion in the area between the two tube passes. The plant personnel routinely ignored replacing the baffle gasket, allowing fluid to leak under the baffle plate at high velocity. The flow velocity scrubbed the corrosion product off the tube sheet, keeping the corrosion rate at a relatively high level and corroding that area at a much higher rate than the rest of the tube sheet. After several years, leaks in the tube sheets required replacement at a tremendous cost.

Comments on Uniform Corrosion

1. Trace contaminants, such as sulfur in an atmosphere, can greatly alter the rate of attack. For example, the areas downwind of major airports tend to suffer greater corrosion rates than the upwind areas.
2. Increases in mechanical stresses can more than double the corrosion rate.

Prevention

Control usually involves either removing the corrodent or the moisture. This could involve chemically preventing oxygen from getting to the surface or using coatings to prevent water from contacting the material. For example, paints and coatings slow the rate at which moisture gets to the metal surface.

Fretting

It is a form of uniform corrosion that occurs when tiny [less than 0.025 mm (0.001 in.)] movements between two parts causes a microscopic welding and tearing of materials.

Appearance

Dark brown or black discoloration between two ferrous parts. (Black is usually an indication that water was present.) Photo 7.4 shows a portion of the impeller fit on a pump shaft. The brownish-black discoloration is fretting and is the result of looseness between the two components.

With stainless steels the fretting often causes a surface that looks polished but is actually slightly pitted and not as smooth as it was initially.

Comments

1. Often seen between a bearing race and the shaft or housing, between a hub and a shaft, and between the strands in a wire rope.
2. False brinelling in bearings is a form of fretting.
3. Fretting will substantially reduce the fatigue strength of the fretted parts.

Prevention

Frequently difficult and depends on preventing movement between the parts. This can be done with tighter fits or adhesives or by changing the design.

PHOTO 7.4 The ugly blackened area on this pump shaft is fretting caused by movement between the impeller and the shaft.

Galvanic Series - in Sea Water (volts)

Magnesium	−1.75
Zinc	−1.10
Al-Zn	−1.05
Aluminum (pure)	−0.8
Mild steel (clean)	−0.5 to .8
Mild steel (rusted)	−0.2 to .5
Cast gray iron	−0.5
Mild steel in concrete	−0.2
Copper, Brass	−0.2
Cast iron (high silicon)	−0.2
Mill scale on steel	−0.2
Graphite	+0.4
Platinum	+0.4

FIGURE 7.11 The Galvanic Series with the metals on top being more active.

PHOTO 7.5 Galvanic corrosion of a steel nipple connected to a bronze valve.

Galvanic Corrosion

This is probably the second most common mechanism of corrosion and happens when two metals or alloys with differing potentials are physically connected and immersed in a conductive liquid. They develop an electrical potential and when the current is allowed to flow, the more active metal (the anode) will be consumed as part of the electrochemical reaction.

Figure 7.11 shows a galvanic series in seawater at ambient temperatures: the metals on the top of the scale, those with the most negative voltages, are more active than those below. So whenever two metals are connected, the more active one corrodes and protects the lower one.

A good example of this is the galvanized (zinc) coating frequently used on steel pipe. The zinc is anodic to the steel it is plated on and protects the steel from corrosion. In a similar manner, the zinc anodes mounted on the wetted area of outboard motors protect the aluminum lower units from corrosion.

Photo 7.5 shows an example of galvanic corrosion. The pieces were taken out of a well 9 months after it had been repiped. On the left is a bronze foot valve that is in like-new condition. To the right of it is a pipe nipple that was originally galvanized. In the middle of the galvanized pipe is a shiny section that is a piece of cellophane tape wrapped around the pipes. The nipple had been in the well for many years while the galvanized coating was doing its duty protecting the steel. However, when the well was repiped, the pipe fitter shortened the nipple and recut the threads on it, exposing clean steel. Unfortunately, after years of use, the galvanizing was no longer there with most of the zinc having corroded to zinc hydroxide, leaving the pipe unprotected. As a result, the assembly surfaces exposed to the water consisted of bronze, freshly cut steel, and old steel. The resulting galvanic corrosion is obvious.

Appearance

Recognition is relatively easy because the corrosion of the more active metal starts right at the junction of the metals, then decreases with distance.

Comments

1. The galvanic series dictates which metal is attacked. The larger the potential difference (voltage) between the metals the greater the rate of corrosion, and the actual rate of metal loss can be predicted based on the current (amperage) flow.
2. Temperature and environment can affect the corrosion rate.
3. The relative sizes of the cathode and anode are critical. A large cathode will generate a large current, and concentrating the current on a small anode will result in rapid loss of material and penetration. This is the *area effect*. (See *Prevention* below.)
4. Whenever new materials are installed in a corrosive environment, the new materials are always anodic to the older ones. (In one installation, we worked on two physically identical pipelines side-by-side. The newer one had five times as many leaks as the older one.)
5. Weld filler metal should have as much or more alloy content than the base metal to avoid galvanic attack.
6. If there are two metals that are electrically connected and the plan is to coat only one of them, be sure it's the cathode. There are always holidays in coatings. If the anode is coated, the corrosion current generated by the cathode will concentrate at a holiday in the anode coating and result in rapid perforation.

Prevention

Prevention usually involves insulating against contact between dissimilar metals. For example, when a new section of buried pipe is installed, an insulating flange kit is installed. This entails using gaskets to separate between the pipe flanges and insulating bushings and insulating washers to ensure there is no conductivity between the two pieces.

However, another approach is to make the anode much larger than the cathode and the area effect is very important in galvanic corrosion. The first galvanic corrosion that we could find reference to occurred in the British navy in the mid-1700s. At that time, the wooden frigate *Alarm* was fitted with copper sheathing to protect the bottom from attack by marine worms. The sheathing was held in position with iron nails, the inevitable galvanic cell was developed, and the nails corroded rapidly. When the ship returned to port with the copper missing, the entire crew was under suspicion of theft and it took some time before they were freed. This is an excellent example of the area effect in which the large cathode generated a current that was applied to the much smaller anode and the loss of material had a rapid effect. If the areas were reversed, with a large anode and a small cathode, not only would the corrosion current be less (and a lesser amount of material removed) but also would be applied over a larger area with a correspondingly lower overall penetration rate.

Photo 7.6 and Photo 7.7 show another example of using the area effect to solve a galvanic corrosion problem. In the first view, we see a galvanized steel nipple where the galvanizing has been exhausted and a hole developed. This nipple was mounted on top of a cast-iron well pump that was constantly moist from condensation and the steel pipe corroded through while it protected the cast-iron housing. The repair consisted of replacing the steel nipple with a copper fitting. Now the cast-iron housing will be the anode, but the corrosion will be so distributed that we will all be long gone before the housing needs replacement.

PHOTO 7.6 This steel nipple was originally in the case iron pump housing shown in Photo 7.7. It corroded as the anode in a galvanic circuit.

PHOTO 7.7 By making the large cast iron part the anode, the assembly life is increased tremendously.

SELECTIVE LEACHING

This is a form of galvanic corrosion in which one component of a metal that is not a true alloy is attacked, leaving the other in place. The most common examples of it are:

- Graphitic corrosion, typically on buried cast-iron pipes, in which the iron is removed and the graphite left behind
- Dezincification of zinc–copper alloys

Appearance

The surprising part of selective leaching examples is that they look almost unchanged from their original appearance, the difference being that one of the components of the material is no longer there. (It has been selectively leached out of the solution.)

Photo 7.8 shows a 20-cm (8-in.) cast-iron fire waterline that had been buried for 50 years when it sprung a leak. When the plant personnel dug up the section, the crack was obvious, but the rest of the pipe looked like new. It wasn't until the piece was sandblasted to remove the graphite that the extent of the attack became apparent.

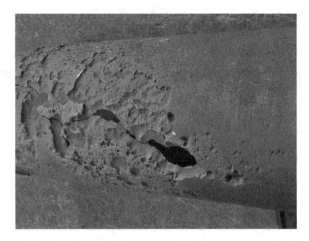

PHOTO 7.8 Selective leaching of a buried pipe.

Comments

1. The term graphitic corrosion is technically incorrect, but it is commonly used to describe the selective leaching of cast-iron pieces.
2. This attack mode is common in buried cast-iron sewer and water pipes. When a section of buried pipe is uncovered, it looks as though there is no damage, but the surface of the pipe can be carved with a sharp knife because the iron is no longer present and part of the pipe wall is essentially graphite. (Cast iron is essentially iron in a graphite matrix and the graphite is the cathode. When very slow-moving water is present, the frequent result is the galvanic attack on the iron and what remains is a graphite pipe.)

Prevention

Selective leaching in buried piping can be prevented with cathodic protection and with coatings, but in most other applications it requires a material change.

Intergranular Corrosion

It is occasionally found in stainless steels but also shows up in other corrosion resistant alloys. It is caused by changes in the metal's chemistry and is usually the result of welding or improper heat-treating procedures.

Appearance

If welding causes it, the most common location for it is the area immediately next to the weld. In other cases, errors in heat treatment can affect entire pieces, and portions of the metal disintegrate and crumble. When welding causes the problem, it will look as though a groove has been machined alongside the weld but not in contact with it. Otherwise the surface will appear rough and feel "sugary."

Comments

1. The mechanism of how it occurs with weldments is relatively complex as follows:
   - The stainless alloy contains both carbon and chromium.
   - When the material is heated above about 525°C (1000°F), the carbon becomes mobile and tends to unite with the chromium at the grain boundaries and form chromium carbides.

- The result is that there is less chromium available to resist corrosion in the areas immediately adjacent to the grain boundary. The chrome-poor section around the boundary then becomes anodic to the rest of the grain, is corroded, and the individual grains fall out.

2. In the past, it commonly happened with corrosive liquids in Type 304 or Type 316 stainless steel. These have been largely replaced with the low-carbon grades, 304L and 316L, which have less carbon available to steal the chromium.

3. The L designation refers to the fact that the amount of carbon is limited. Type 304 allows as much as 0.08%, whereas 304L limits the carbon content to 0.03%. As a result, fewer chromium carbides are formed at the grain boundaries and the rate of attack is greatly reduced.

4. The downside to the reduced carbon level is that the L grades are somewhat weaker than the higher carbon alloys, and some of them have other alloying materials, such as nitrogen, to increase their strength.

5. There have been instances of intergranular corrosion of aluminum alloys and brasses.

Prevention

With weldments, the most common approach is to use low-carbon material (304L, 316L) so there won't be carbon available to unite with the chrome. However, there is still some chrome available, and the net effect is that the rate of intergranular corrosion is greatly slowed but not stopped. If total prevention is needed, the best approaches involve altered heat treatment and modified alloys.

In some cases, intergranular corrosion involves attack such as that shown in Photo 7.9. This is a stainless steel pump impeller that was improperly heat treated. When exposed to concentrated phosphoric acid, the grain boundaries were attacked and the grain structure can be clearly seen in the photo.

EROSION-CORROSION

As mentioned earlier in this chapter, as the flow rate of a corrodent increases the corrosion rate almost always also increases. This is called *flow accelerated corrosion (FAC)*. However, the difference between FAC and erosion-corrosion is that FAC tends to scrub the surface clean and expose it to fresh corrodent, whereas erosion-corrosion is a result of a combination of high-velocity mechanical removal of an oxide scale and corrosion, thereby greatly increasing the loss rate. Erosion-corrosion frequently happens when there are solids in the liquid.

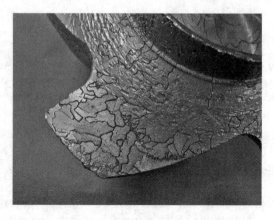

PHOTO 7.9 Intergranular corrosion of a pump impeller.

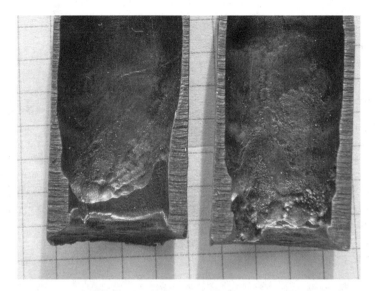

PHOTO 7.10 Erosion corrosion of two heat exchanger tubes.

PHOTO 7.11 Erosion corrosion of a pump impeller.

Appearance

Appears as smooth-bottomed pits, frequently either horseshoe-shaped or superimposed on one another, with a directional pattern related to path of fluid. The top edge of the pit is frequently sharp and pointed downstream. Photo 7.10 shows a pair of 2.5-cm in. tubes that have been removed from an ammonia vaporizer and cut in half. The liquid ammonia comes in the bottom of the tubes and flashes into a gas. There is some deformation on the lower edges of the tubes from the removal tools, but just above that an area can be seen where the wall has been substantially thinned. The tube to the left shows more classical symptoms of erosion, whereas the tube to the right shows obvious erosion and corrosion attack.

Another example of erosion-corrosion is shown in Photo 7.11 where a piece of a pump impeller shows a very definite flow pattern from the left to the upper right.

Comments

Erosion-corrosion thrives on high-velocity conditions and occurs in ells, tees, valves, pump cases, impellers, and other places with high velocity and turbulence.

For erosion-corrosion to take place, there has to be a corrosion mechanism at work.

Prevention

Reducing the flow rate, changing to a harder material, or changing to a more corrosion-resistant material are the methods of prevention. In cases where there are solids in the flow stream, eliminate the solids or reduce their size.

> The most amazing erosion-corrosion we ever saw was on a boiler tube in a sulfur-burning boiler. The plant had been having problems with tube failures, and we had just completed an extensive series of thickness tests on the tubes. We had found that some were thin, and the plant put the boiler back on line with the idea that they would change them in the near future. Eventually, one of the thin tubes failed and blew hot water and steam into the burning sulfur stream, washing across the adjacent tubes. Then a second tube failed but it wasn't one of the thin tubes, and we had recorded its thickness at 5.1 mm [0.20 in.]. The high-velocity hot-acid erosion-corrosion attacked the tube at a rate of approximately 178 m/year [584 ft/year].

Cavitation

Cavitation is a form of erosion-corrosion in which vapor bubbles are formed and their collapse causes tremendous impact forces. It is usually found in pumps but can occur anywhere that the forces cause vapor bubble development and collapse.

Appearance

It frequently looks similar to fine pitting. On pumps, as shown in Figure 7.12, starvation cavitation damage occurs on the backside (negative pressure side) of the vanes near the eye of the impeller while recirculation cavitation appears on the pressure side of the vane near the OD and sometimes on the back of the impeller. Microscopic or metallographic examination is used to identify the damage, and the surface grain plastic deformation can be plainly seen under high magnification.

On comparing the two cavitation mechanisms in pumps, there are some very obvious differences. The figure shows how the locations differ but in addition to that:

- Starvation cavitation is caused by a lack of pressure on the fluid at the eye of the impeller. The fluid vaporizes and then the vapor bubbles collapse as they progress into higher pressure areas. The result is a noisy pump that sounds like it is pumping nuts and bolts, and the forces are tremendous and can destroy the pump bearings in less than a day.
- Recirculation cavitation is caused by the pump's inability to properly discharge the fluid. When the discharge is throttled, as the vane passage goes past the high-pressure area of

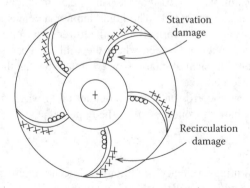

FIGURE 7.12 Location of cavitation damage on a pump impeller.

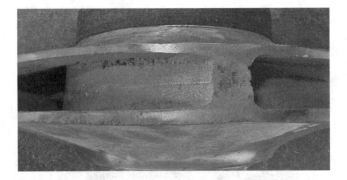

PHOTO 7.12 Recirculation cavitation damage on a pump imepller.

the volute, some of the liquid is forced back down the passage. In severe cases recirculation cavitation can destroy an impeller in less than a year.

Photo 7.12 shows a view of a pump impeller with holes alongside the passage and at the end of the vane that were caused by recirculation cavitation.

Comments

1. Cavitation forces can be as high as 550 MPa (80,000 psi.)
2. Austenitic stainless steels and aluminum alloys are particularly vulnerable at high flow velocities.

Prevention

Three approaches to prevention are the following:

1. Improve the flow stream by eliminating the cavitation action.
2. Select higher strength alloys or hardfacing alloys in which the surface is stronger than the impact forces from the cavitation bubbles.
3. Use corrosion resistant alloys to remove one portion of the erosion-corrosion action and greatly improve component life.

In one example, there was a 250-hp pump and recirculation cavitation had completely destroyed the impeller in less than a year. It pumped water from a local stream that was used by the plant for cooling. The impeller was changed to a hardened high-chrome cast-iron one, and an inspection after 6 months of operation found the original machining marks were still visible. The pump was shut down 10 years later, and the impeller was still in good condition.

CONCENTRATION CELL CORROSION

It is caused by differences in concentration in the solution contacting the target material. These concentration differences cause electrical potential differences and a corrosion cell begins.

Appearance

This relatively common form of corrosion is often seen at the waterline of a liquid and on buried materials. Photo 7.13 shows four structural steel bolts where the concentration cells at the bolt heads started the corrosion that has begun to destroy the surrounding paint.

PHOTO 7.13 Concentration cell corrosion is starting to lift the paint on this structure.

Comments

1. Similar conditions cause oxygen differential cell formation under deposits or in other areas deprived of oxygen, i.e., between dirt, corrosion products, or scale and metal wall.
2. The difference in available oxygen (an oxygen differential cell) contributes to most buried pipe leaks occurring at the bottom of the pipe.

Prevention

It can be combated by changing materials, but a common approach on tanks and pipelines is cathodic protection.

Crevice Corrosion

This is a very localized concentration cell mechanism that occurs in crevices wide enough to allow liquid to enter, but so narrow the liquid is stagnant. Cracks of about 0.1 mm (0.004 in.) are just right.

Appearance

Crevice corrosion is found under rivet heads, bolt heads, nuts, lap joints, incomplete weld joints, area between gasket and flange faces, absorbent gaskets, etc. The attack shown in Photo 7.14 started at the surface of this Type 304 SS instrument tube that was immediately adjacent to another piece. When the adjacent piece was removed the attack was obvious.

PHOTO 7.14 Crevice corrosion damage on a 304 stainless tube that was up against another piece of type 304 stainless.

PHOTO 7.15 Pitting corrosion adjacent to the weld on a steel elevator cylinder — caused by the residual stress.

Comments

1. Often occurs in fluids containing chlorides.
2. Austenitic stainless steels are particularly susceptible.

Prevention

Eliminate crevices and deposits, change material, and apply coatings. (This sounds like a good correction practice, but it is much easier said than done.)

PITTING CORROSION

Pitting frequently appears as localized, small, almost random holes in an otherwise unaffected area and usually occurs in materials that have a protective coating. Holes can be deep or even through the wall. Pitting frequently acts as a stress concentration factor, starting fatigue cracks, and is sometimes difficult to detect as the pits frequently start under deposits and are covered by the corrosion products.

Appearance

There are two general types of pits:

1. In carbon steels they are wide and shallow and progress at a very slow rate. Photo 7.15 shows some pits in a mild-steel hydraulic elevator cylinder. The cylinder was buried in a wet soil and wasn't properly coated and wrapped, so the water leaked in around the weld. The pits started here, because the residual stress from the welding made the adjacent parent metal anodic.
2. In stainless steels the pits often grow bigger as they progress through the part, and the entrance may even be minute.

Photo 7.16 shows two aluminum housing covers that were separated by about 20 cm (8 in.). The solution running through the housing was mildly corrosive to aluminum, and the plant had put a condensate hose on the left cover to prevent the product from freezing in the housing. The increase in temperature resulting in a good example of Arrhenius' rule, and Photo 7.17 is a close-up of the heavily pitted surface.

Comments

1. Pitting occurs commonly in austenitic stainless steels, but also is found in aluminum and carbon steel (and other metals that rely on protective oxide coatings), almost always in stagnant areas that allow the localized corrosion cell to develop. The presence of a liquid is a necessity, and we have seen it happen at temperatures over 600°C (1100°F).

PHOTO 7.16 Pitting corrosion on one of two aluminum castings.

PHOTO 7.17 The heavily pitted surface caused by elevated temperatures.

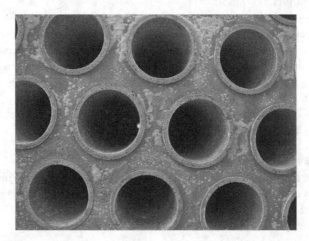

PHOTO 7.18 The pitted surface of a heat exchanger. In this case the pitting was caused by increased solution concentration.

2. Cause — Local cell develops and breaks through the protective (oxide) film. In stainless steels, the combined presence of soil deposits and chlorides frequently accelerates the process. Photo 7.18 shows the tube sheet of a Type 316 stainless steel heat exchanger. The plant personnel wanted to accelerate the cleaning of the heat exchanger bundle, so

PHOTO 7.19 Pitting corrosion of a weld that was slightly anodic to the parent material.

they intentionally increased the chloride concentration in the cleaning solution. A serious problem lies in the fact that pitting is very difficult to stop, because the inside of the pit is anodic to the surroundings and oxygen can't readily enter the pit to reestablish the oxide coating. In this example, they ended up with holes in several of the tubes.

3. Pitting can be a rapid process. 0.5 mm/month (0.02 in./month) is not uncommon in stainless steels.

Prevention

1. Use of pitting-resistant metals. In piping and tanks, increased flow rates and/or frequent cleaning can prevent deposits from forming and can prevent chlorides from concentrating, both of which help prevent pitting from starting.
2. As mentioned earlier, correction is almost impossible. Once a pit starts, the anodic action within the pit interior acts to continue the pitting process.
3. Photo 7.19 shows pitting along the edges of a stainless steel weld in a stainless duct. The duct carried acid gases that were occasionally condensing. The weld metal was ostensibly the same as the duct alloy but in reality it had a little less chrome — so it became anodic. The oxide film prevented uniform attack but couldn't stop the pitting in the highly stressed areas. A good practice is to always ensure that the weld metal is more corrosion resistant than the parent metal.

STRESS-CORROSION CRACKING (SCC)

Almost every presentation on SCC includes a discussion on the "SCC triangle" shown in Figure 7.13. As shown in the diagram, in order for SCC to occur there has to be sufficient stress, the material has to be sensitive to the atmosphere, and the correct atmosphere has to be present. An excellent example of the need for all three can be seen in a milk tank inspection project that we worked on. For many years we inspected the welds in a series of Type 304 SS tanks (shown in Photo 7.20) to ensure there were no cracks, and for many years we found nothing of concern. Then one year we found a small crack in one of the welds. The plant people said nothing had changed, but the next year we found several cracks and had lengthy meetings to determine why they had appeared. Eventually we found out that they had increased the temperature of the water used to clean the tanks. The cracks began on the exterior of the tanks. The atmosphere where the cracking

FIGURE 7.13 The stress corrosion cracking triangle. All three are needed for SCC to occur.

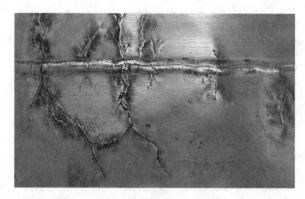

PHOTO 7.20 SCC on the outside of a milk storage tank.

PHOTO 7.21 SCC — these jagged, irregular cracks indicate that there is a metallurgical incompatability.

started hadn't changed and the metal certainly hadn't changed, but the increased temperature increased the stress in the welds, and SCC appeared.

Appearance

These are ragged, irregular fine cracks with multiple branches while the rest of the piece looks absolutely unaffected. The cracks frequently start at pits. Photo 7.21 shows a 1.2-cm (1/2-in.) tube with a lengthy jagged and disjointed stress corrosion crack.

Comments

1. The amount of stress can be small, as low as 14 MPa (2000 psi), but it can also range up to almost the yield strength of the material.
2. Discontinuities such as pits, gouges, and deep scratches are frequently starting points. Photo 7.22 shows a view of a small-diameter instrument tube as seen through a microscope. The stress-corrosion crack is obvious and a close examination can see the pits where it started.
3. It can sometimes be found at ambient temperatures, but becomes much more common at higher temperatures. We have seen it in equipment that never saw temperatures above 25°C (75°F), but it becomes much more common above 50°C (122°F), and there are people (and books) that say it does not occur below 50°C.
4. Troublesome environments — Caustic soda for inconel, monel, and nickel; chloride and fluoride solutions for austenitic stainless steels; caustic and nitrate solutions for carbon steel; ammonia compounds with copper alloys. All metals have some condition that causes SCC.
5. Caustic embrittlement is a form of SCC that happens when carbon steels are exposed to caustics, and it generally occurs at elevated (over 50°C) temperatures but can happen at much lower temperatures.

PHOTO 7.22 The SCC cracks started from a series of small pits.

Prevention and Correction

1. Prevention — Remove the stress, change the atmosphere, or the material. (Sometimes decreasing the temperature can help.)
2. Once the material has cracked, it is almost impossible to stop further cracking. This is a corrosion mechanism so elimination of all moisture should effectively stop the crack growth progression, but this is almost impossible to accomplish. However, if the stress causing the cracking decreases, the cracking will stop. (For example, if the residual welding stress is the source of the stress, the cracking will decrease as the field stress decreases.)
3. One serious limitation of materials that have visible SCC is that there are many more microscopic cracks. The SCC cracks will grow in a predictable manner, but the microscopic cracks will have seriously reduced the toughness of the material. Stress-corrosion cracked components can be used but it is important that the limitations be recognized and cautions applied.

HYDROGEN DAMAGE

This description covers a multitude of problems and is described in sometimes confusing terms. Understanding the hydrogen damage mechanisms starts with the basic corrosion mechanism. Atomic hydrogen is generated at the cathode, and most of the hydrogen atoms join with other atoms to form the familiar molecular hydrogen gas, H_2. But some of the atoms remain unattached and they are much smaller than the voids between the iron and steel atoms. These hydrogen atoms enter the metal and travel through it causing a variety of problems including hydrogen embrittlement, hydrogen-influenced cracking (HIC), and hydrogen blistering.

Hydrogen Blistering

This is the easiest to understand of all the mechanisms. When a free hydrogen atom migrates through a metal and finds another hydrogen atom, the two atoms unite to form molecular hydrogen, H_2. The unifying forces that are developed when the two form this molecule are much stronger than the metal, and a blister develops inside the metal wall. Photo 7.23 shows a small plate with some of these blisters.

Appearance

These blisters range in size from that of a small pea to the size of a football. It frequently happens at elevated temperatures and pressures, where there is substantial dissociation of the hydrogen molecules, but can occur at ambient temperatures.

Prevention

Correction is difficult and action is most often directed toward monitoring. However, inhibitors and coatings have been used successfully.

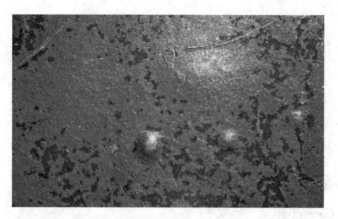

PHOTO 7.23 Hydrogen blisters in a mild steel plate.

Comment

Most of these occur during hydrocarbon processing but large carbon steel sulfuric acid storages are also a common location for hydrogen blisters. In the acid storages, the hydrogen is generated by the reaction that forms the ferrous sulfate layer and tends to collect in areas. Any welding or torch-cutting of these acid tanks and lines is frequently accompanied by sputtering and popping of many small blisters.

Hydrogen-Influenced Cracking (HIC) (Also Called Hydrogen-Induced Cracking)

According to current thinking, the mechanism is as follows:

- Hydrogen migrates through a metal and unites with other materials to form hard, brittle compounds.
- The metal loses ductility in the area of these intermetallics and elevated pressures are present.
- Cracking then can occur at loads well below the tensile strength.

It is commonly seen in higher strength (HRC 32 and above) steels near processes that liberate hydrogen but can occur in lower strength steels in more aggressive and corrosive atmospheres.

Appearance

The appearance is essentially indistinguishable from SCC with ragged, disjointed cracks as seen in Photo 7.24. This is a cross section of an ASTM A 193 Grade B7 bolt that was exposed to a mildly corrosive atmosphere. The bolt is being wet fluorescent magnetic particle tested and, using a black light, the cracks show up as white lines.

This mechanism can also appear as seen in Photo 7.25. This is a spring with a corroded mounting surface. The end coil was ground flat and it rested on a plate that was frequently wet. Eventually, some corrosion occurred, and the generated hydrogen combined with the high hardness of the spring resulting in several failures. This is a photo made through a 7× microscope and the semi-circular original crack is plainly visible.

Comments

1. Any process that generates hydrogen, i.e., corrosion, plating, welding, and other industrial processes, can cause these problems.
2. Higher-strength steels are much more sensitive to the presence of atomic hydrogen.
3. With welding, this is the reason why low-hydrogen rods are beneficial.

PHOTO 7.24 HIC cracks in a light high-strength bolt.

PHOTO 7.25 Hydrogen initiated the crack in this spring.

4. Appearance is typically a jagged crack with fine branching cracks adjacent to the primary one. (Similar to SCC.)
5. There is confusing terminology about what we have called HIC. In the fastener industry, this mechanism is known as *hydrogen embrittlement*, whereas most corrosion engineers and scientists call it *HIC*.
6. The presence of sulfur compounds aggrevates the problem by preventing or slowing conversion to molecular hydrogen.

Prevention

• In manufacturing processes, if the hydrogen source cannot be eliminated, the parts are baked at 190°C (375°F) for 23 h to drive out the hydrogen. (In order to prevent permanent internal damage to the parts, this has to be done in a relatively short time after the hydrogen is absorbed.)
• In plant operations involving corrosion, the materials have to be changed to eliminate the problem.

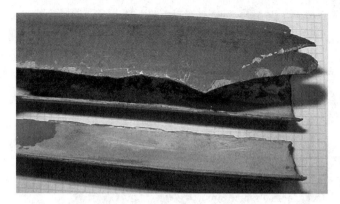

PHOTO 7.26 True hydrogen embrittlement where the outer portion of this titanium tube has become titanium hydroxide.

Hydrogen Embrittlement

In the eyes of corrosion engineers and scientists, there are two forms of hydrogen embrittlement:

1. The metal absorbs hydrogen and is converted to a brittle nonmetallic compound. For example, Photo 7.26 shows a piece of titanium tubing that is about 2 cm (1 in.) in diameter. Looking at the right side of the photo, it can be seen that there are chips out of the tube that appear to be classical brittle fractures. Yet the portion of the tube that was not in the gas stream is still as ductile as a paper clip. The tube was exposed to steam and the free hydrogen atoms have converted the external portion of it to titanium hydride, and it is almost as brittle as glass.
2. The other embrittlement mechanism involves the hydrogen atoms meeting scattered nonmetallic ions within the metal and forming compounds that result in internal pressures that greatly reduce the tensile strength of the metal.

A Special Category: Microbiologically Influenced Corrosion (MIC)

MIC deserves special mention because of its effect in causing corrosion. Figure 7.10 shows the eight corrosion categories, and three of them, uniform corrosion, pitting, and crevice corrosion, are sometimes affected by microbial action. Twenty-five years ago, many corrosion specialists thought that perhaps 5 or at most 10% of the corrosion was caused by MIC. Today the consensus is that a substantial percentage, perhaps as much as 40% of all corrosion, is influenced by these living organisms. The organisms don't actually do the corroding, they accelerate the corrosion. A good example of this was the brine lines that ran from our brine wells to the plant. There were microbes that lived in colonies along the pipe wall. The microbes consumed the sulfur compounds that were in the brine and secreted sulfuric acid, which didn't help the pipe at all.

The situations range from deposits caused by slimes that tend to be resident sites for crevice corrosion to microbe colonies that physically consume metals. There are a multitude of possible situations, and testing is recommended if MIC is suspected at all.

- Identification is difficult because microbes are frequently present but may not be involved with the corrosion. Three sources of test kits used to detect the presence of MIC are Bioindustrial Technologies Inc., DuPont, and Hoch.
- Environment — Liquids must be present. Observed temperatures range from 0 to 110°C (32 to 230°F). Observed pH range is from 0 to 14. Normally it does not occur in flows greater than 1.5 m/sec (5 ft/sec).
- Combating MIC depends on the specifics of the case. Alternatives include biocides, process changes, and material changes.

8 Lubrication and Wear

In understanding the effects of various lubricants and how they affect machine life, it makes good sense to first look at some simplified versions of how the basic bearing types work, then review how lubricants are formed, and then return for a more detailed look at various applications and how lubricants affect them.

THREE TYPES OF LUBRICATED CONTACTS

There are three basic types of lubricated contacts:

1. Rolling element (ball and roller) bearings such as those seen in most common industrial motors, reducers, and pumps.
2. Plain (sleeve) bearings as seen in large motors, generators, and the main and rod bearings in most of our automobile engines.
3. Sliding bearings in which pieces are in contact with and sliding over each other, i.e., a trailer hitch or slider plate.

In looking at these bearings, we see they involve three very different lubrication regimes that really require very different oils and greases to perform their jobs most efficiently.

LAMBDA: THE LUBRICANT FILM THICKNESS

Probably the first concept to understand is that of the λ ratio (pronounced "lam da"). This is the thickness of the lubricant film and is a comparative measurement of the separation between two pieces and their relative roughness. As the separation between the two pieces increases, λ increases and the wear rate decreases, as shown in Figure 8.1.

This film thickness is the result of the lubricant viscosity, the difference in speeds between the two pieces, and the shape of the parts, and it typically ranges from 0 to 4. The fluid's viscosity is important because it is a measurement of how readily the liquid flows when subjected to a load. High-viscosity fluids tend to resist and stay in position, whereas low-viscosity fluids are easily displaced. In Figure 8.1, the right half of the chart is largely controlled by the lubricant's viscosity, whereas the left side involves a complex relationship between the oil, any thickener, and the additives.

Using these values and the separation between the pieces as a guide, there are three general classifications for categorizing lubrication regimes. They are:

1. *Boundary lubrication* — This involves a range of lubricant film thicknesses in which the moving components are not totally separated from each other. It almost always occurs with slow speeds and frequently involves grease lubrication of parts slowly sliding over each other. In this area, the additives are usually the most important components of the lubricant, and the properties of the lubricated mating surfaces determine the friction and wear rates. Typical examples include slow-moving plain bearings and heavily loaded gears.

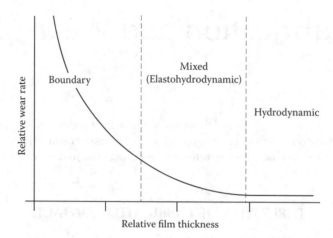

FIGURE 8.1 The three basic lubrication regimes with greatly decreasing wear rates as the film thickness increases.

2. *Mixed-film* (or *elastohydrodynamic*) *lubrication* is used in areas where ball and roller bearings and high-speed gears are found. Within this area, keys to the wear rate are the viscosity, cleanliness of the lubricant, and deformation of the mating surfaces.
3. *Hydrodynamic lubrication*, also called full-film lubrication is that area where the combination of lubricant and operating properties results in true separation of the pieces by the liquid and very low wear rates result. It is in this range that minimum wear and maximum efficiency take place.

BASIC BEARING LUBRICATION

Elastohydrodynamic

This is also frequently called *mixed-film lubrication* and occurs when the rolling elements of ball and roller bearings travel across a lubricated plane. In Figure 8.2 we see the side view of a typical rolling element bearing. At moderate to high speeds, at the incoming edge of the rolling contact, the lubricant inlet zone, the lubricant film is trapped and viscosity conversion takes place. At this point, the easy flowing liquid, the oil, is transformed to a semiplastic fluid with a viscosity that is a hundred or even a thousand times greater than that same oil under atmospheric pressure.

Looking at the Hertzian contact region of this rolling element bearing, we see that the change in lubricant viscosity results in a separating film between the flattened portion of the element and the plane it is in contact with, greatly reducing the local stress and increasing the life of the bearing.

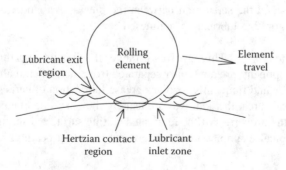

FIGURE 8.2 The rolling element traps lubricant in the inlet zone and the extraordinary pressure of the Hertzian contact zone increases the viscosity tremendously.

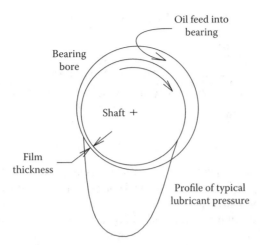

FIGURE 8.3 The layout of a typical plain bearing showing the lubricant pressure profile.

Then, as the lubricant leaves the contact region, it instantly converts back to its unpressurized viscosity. The film thickness in this bearing is generally in the range of 0.0005 to 0.001 mm (0.00002 to 0.00004 in.), which is about 1/100th the thickness of a sheet of textbook paper, and the pressure is on the order of 2 GPa ($\approx$ 300,000 psi). (The term *elastohydrodynamic* is used because the lubrication mechanism involves both elastic deformation of the rolling element and hydrodynamic action of the lubricant.)

Hydrodynamic Action

The difference in action between a rolling element bearing and a plain bearing is dramatic. The hydrodynamically lubricated plain bearing has much lower pressures, rarely exceeding 15 MPa ($\approx$ 2200 psi), the clearances are on the order of 0.01 to 0.05 mm ($\approx$ 0.0004 to 0.002 in.), and there is no viscosity conversion. As shown in Figure 8.3, oil is fed into the bearing and trapped between the rotating shaft and the fixed outer bore. Primarily a result of the lubricant viscosity, the trapped oil can't escape, and it provides a low-friction cushion for the shaft to operate on.

Boundary Lubrication

The third type of lubricated contact involves very slow rolling element bearings and sliding contact between mating pieces. In the previous two examples, the viscosity and cleanliness of the oil were critical. In this example, the lubricant additives provide the operating surfaces that resist the wear, and Figure 8.4 is a simplified view of what happens.

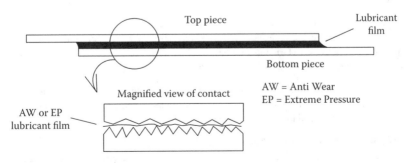

FIGURE 8.4 In a boundary lubricant the film thickkness approaches zero and the additives are the keys to reducing wear.

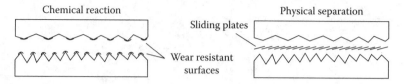

Magnified views of contact surfaces with EP additives

FIGURE 8.5

The action of the additives is critical to the effectiveness of these applications. Antiwear (AW) lubricants have a high proportion of polar fatty acid compounds that in operation tend to be attracted to the metal surfaces. These molecules line each side of the gap, helping to prevent contact between the opposing asperities. Extreme pressure (EP) lubricants have far more load-carrying ability than AW lubricants and function in the two ways shown in Figure 8.5. Explaining these two methods:

1. Physical separation involves the use of additives such as molybdenum disulfide, graphite, and PTFE. In preventing wear, the additives provide cushioning and physical separation between the two pieces.
2. With chemical action, the additives react to the heat and pressure generated by contact and develop coatings using the sulfur, chlorine, and phosphorous additive compounds to provide wear-resistant surfaces.

Most EP greases and oils combine a variety of additives to help perform their duties over a wide range of operating conditions.

Summarizing the bearing characteristics, Table 8.1 is a brief chart comparing the differences between three basic types of lubrication for contact surfaces.

TABLE 8.1

Type of Contact	Typical Max Contact Pressures (MPa/psi)	Typical Film Thickness (mm/in.)	Effect of Water
Rolling Element	2 GPa (300,000 psi)	0.00075 mm/0.0003 in.	0.4% cuts life by 5 (Note 3)
Plain	15 MPa/2,200 psi	0.03 mm/0.001 in.	Note 1
Sliding	1/2 metal's yield strength	0	Note 2

Notes:

1. From our discussions with technical lubrication personnel and equipment builders, the primary problem caused by water contamination of plain bearing lubricants is that their corrosion resistance is reduced. The consensus of the folks we have talked with is that water contamination becomes harmful when there is more than 1.5% water in the oil.

2. The primary problem with sliding bearing lubrication and water involves the reaction with EP additives. AW additives are essentially immune to deterioration from water, but EP additives such as sulfur and chlorine tend to form acids as they degrade.

3. We will further discuss this in Chapter 10.

MANUFACTURING A LUBRICANT

VISCOSITY MEASUREMENT AND VISCOSITY INDEX

The base oil is the primary raw material that the lubricants are made from, and the viscosity at the operating temperature is critical for most applications. The following is a review of lubricant viscosities and how they change with temperature.

There are several units for measuring viscosity, and the most common one used in North America up until about the mid-1970s was the Saybolt Universal Second (SUS). It has since been replaced by the metric centistokes (cSt) but the SUS is a good place to start because it is so easy to understand.

Figure 8.6 is a chart with the viscosity ranges as the vertical scales, and the Saybolt scale is on the left side. To check a lubricant's viscosity using a Saybolt viscometer, a small sample of the liquid is put into the sample cup and heated. In the viscometer, there is a mixer to ensure the temperature of the oil surrounding the cup is uniform, and a very accurate thermometer. When the liquid in the cup gets up to the test temperature, the standard Saybolt orifice in the bottom of the cup is opened, and a stopwatch is started. The liquid starts to drain into a small beaker, and when the level in the beaker reaches the 60-ml line, the watch is stopped and, if it takes 300 sec to fill the beaker to the line, the viscosity is 300 SUS.

Looking at the figure, at a temperature of 20°C (68°F), the viscosity of automotive antifreeze is about 50 SUS, and honey is somewhere between about 5,000 and 10,000 SUS. (The system for measuring the kinematic viscosity of fluids using centistokes is similar and well proven but it doesn't have the simplicity of the SUS approach. A fairly accurate conversion factor that can be used between about 30 and 2,000 cSt is: cSt × 4.65 = SUS.)

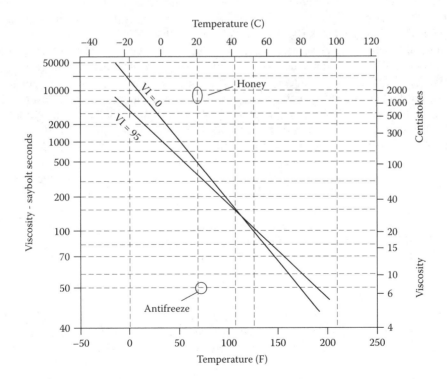

FIGURE 8.6 Showing how a lubricant with a VI of 0 thins more rapidly than one with a VI of 95.

Crude Oil Source	Typical VI Range
Pennsylvania	90 – 110
North Sea	55 – 70
Texas	25 – 55
Alaska's North Slope	15 – 22
California	5 – 15

FIGURE 8.7 The VIs of these major crude sources vary tremendously.

VISCOSITY INDEX

Looking at the two lines in Figure 8.6, it is obvious that their viscosity changes with temperature and that they don't change at the same rate. These two lines show the viscosity of two SAE 10 (or ISO viscosity grade 32) oils, and the difference in the way they thin is because they have different viscosity indices.

About 80 years ago, as automobile engines and their lubrication became more sophisticated, it was realized that, as temperatures increase, some oils lose their viscosity more rapidly than others. In response to this, a measurement system was devised to show the differences in this rate of viscosity change, and it was called the *viscosity index* (VI). Figure 8.7 shows the VIs for five common crude oil sources with the higher VI oils thinning less rapidly as temperatures increase. This difference in VIs is a good part of the reason why early Pennsylvania-based crudes with their paraffinic base oils developed their reputation for being good engine oils. Today, most lubricating oils are still based on paraffinic crudes and almost all of the base oils have VIs of 95 or greater. In addition, there are many motor oils using synthesized base stocks that have relatively high VIs, and some of the polyalphaolefins (PAOs) have VIs greater than 170.

The classification of engine oil viscosities, i.e., SAE 10, 20, etc., is still universally measured by the system used in SAE standard J 300a, and the measurement of automotive gear oils, SAE 75W, 85W, 90W, and 140W, is usually done by SAE standard J 306a. Most of the other oils we use follow the ISO classification outlined in ISO 3448. This standard has 18 viscosity grades as follows: 2, 3, 5, 7, 10, 15, 22, 32, 46, 68, 100, 150, 220, 320, 460, 680, 1000, and 1500. Except for the first four grades (2 to 7), the ISO viscosity grade number is the midpoint of the kinematic viscosity at 40°C (104°F) and the tolerance is ±10%, so the viscosity of an ISO 32 oil would be 32 cSt ±10% or 28.8 cSt to 32.2 cSt at 40°C (104°F).

Figure 8.8 puts much of what was covered earlier together in a chart that shows the approximate boundaries for both the SAE and ISO viscosity classification systems and how they vary with temperature.

The oils shown in this chart have a VI of 100 but there are many multiviscosity oils used primarily for engines and hydraulic systems. In the case of multiviscosity motor oils, an SAE 10W-30 motor would act as a 10-weight oil at −18°C (0°F) and a 30-weight oil at 99°C (210°F).

ADDITIVES

Regardless of the lubricant's viscosity, if it doesn't have the necessary chemical and physical materials combined with it, it won't be able do the job. Additives are what differentiate an automotive engine oil from a diesel engine oil and a steam engine oil and they are critical to the lubricant's function.

An *additive* is a material added to a base oil or grease. Most additives improve the performance characteristics of the lubricant but some, such as dyes and perfumes, are used to change the user's

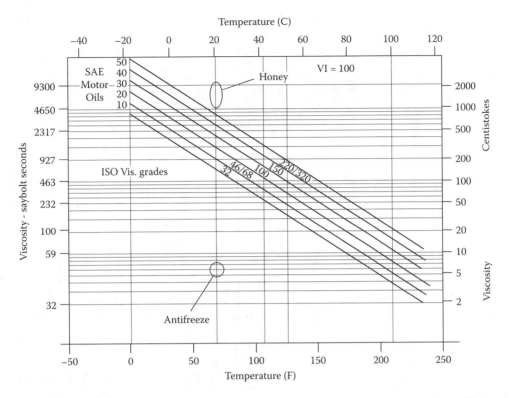

FIGURE 8.8 A viscosity vs. temperature graph showing some of the more common automotive and industrial oils.

perception of the lubricant. Figure 8.9 shows the formulation of a motor oil, and all finished oils are combinations of a base oil and several additives. Greases are the combination of a base oil, additives, and a thickener. (In the section in which we talked about the composition of steels, we mentioned that making a steel alloy was a little like making a cake, where certain ingredients are added to get a specific product. The same basic approach holds true for making a lubricant.)

Some of the more common additives are:

- *Oxidation inhibitors* — Reduce the formation of gums, sludges, and acids. Usually, they are sulfur or nitrogen compounds but other chemicals such as phenols and amines are also used. Oxidation inhibitors are found in most oils and greases except those with short expected lives.
- *Viscosity index improvers* — Reduce the effect that temperature has on the fluidity of an oil. VI improvers reduce thinning at high temperatures and thickening at low temperatures. Normally, polymers such as polyisobutane are used. VI improvers are found in many engine and multiviscosity hydraulic oils. (In making a conventional engine oil with a group II base oil, the base oil is usually at low viscosity, i.e., SAE 10 or so, then VI improvers are added, so it behaves as though it were a 30- or 40-weight oil at high temperatures. The base oils used in synthetic and group III motor oils have higher VIs than the group II oils and don't require as much VI improver.)
- *Pour point depressants* — Prevent wax crystals from jelling at low temperatures, allowing the oil to flow more freely. These are usually aluminum soaps and aliphatic petroleum fractions. Found in oils that are expected to be used over a wide range of temperatures such as automotive, hydraulic, and some low-viscosity gear oils.

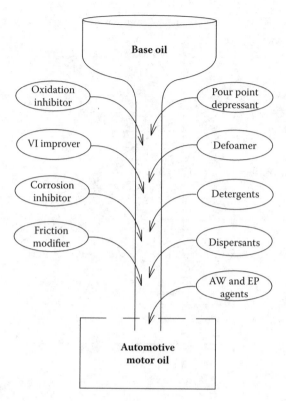

FIGURE 8.9 Making an automotive engine oil.

- *Defoamers* — Minimize the formation of foams by separating the bubbles from the oil. They are critical for engine oils and hydraulic fluids and are found in every type of oil subjected to mixing with air. The most common defoamers are liquid silicones (poly-dimethylsiloxanes).
- *Rust and corrosion inhibitors* — Reduce the probability of corrosion taking place in the lubricated areas. These vary from simple compounds to complex anionic polar film-forming additives that prevent water from reaching the metal surfaces.
- *Emulsifiers and demulsifiers* — Depending on the specific additive, they either promote or prevent the miscibility of water with the lubricant. Demulsifiers are critical for many applications, such as turbines, where the separation of water and lubricant is important. In other applications, such as low-speed wheel bearings that may be contaminated with water, emulsifiers are valuable in preventing corrosion and maintaining the lubricant action.
- *Antiwear additives* — As mentioned above, these are generally polar fatty acid compounds that adhere to the opposing metallic surfaces, providing tenacious low friction separating layers between the surfaces.
- *Extreme pressure* — EP oils are more effective in preventing wear than antiwear oils but are generally shorter lived. They are found in gear and hydraulic oils and there are two general types of EP additives.
- *Detergents and dispersants* — Detergents allow the oils to clean deposits from the surfaces the oils come in contact with. Dispersants prevent deposits from initially adhering to those surfaces. In combination, the two additives keep contaminants in suspension

and allow filtering systems to act more effectively. They are frequently found in engine oils and hydraulic oils.

- *Friction modifiers* — Additives that reduce internal lubricant friction and provide for better mileage. They are found in most automotive engine oils.

In the latest issue of the AISE *Lubrication Engineers Manual*, there are 64 different lubricant classifications (grades). The differences involve variations in both base oils and their additives and Table 8.2 shows the customary additive applications.

Recognizing the important effect that base stocks have on finished engine oils, the American Petroleum Institute has divided lubricant base stocks into five categories:

- Group I includes base stocks that contain less than 90% saturates; and/or more than 0.03% sulfur; and have a viscosity index at or above 80, but below 120.
- Group II includes base stocks that contain 90% saturates or greater; and/or only 0.03% sulfur or less; and have a viscosity index at or above 80, but below 120.
- Group III includes base stocks that contain 90% saturates or greater; and/or 0.03% sulfur or less; and with a viscosity index of 120 or higher.
- Group IV includes PAOs, the most popular synthetic base stocks for formulating automotive engine oils.
- Group V includes all other base stocks not included in the first four groups.

Most base oils used for motor oils prior to 2001 were Group II. With the increased VI requirements of the newer automotive engine oils, most of the base oils are now Group III and Group IV. Gear lubricants are mostly Group II, although there is increased penetration by Group III and Group IV.

TABLE 8.2

Additive Category Lubricant	Oxidation Inhibitor	VI Improver	Pour Point Depressant	Rust and Corrosion Inhibitor	Defoamer	AW/ EP	Detergent	Dispersant	Dye
Automotive engine oil	X	X	X	X	X	X	X	X	S
Automatic transmission fluid	X	X	X		X	X	X	X	X
Industrial gear oil	X	S	S	X	X	X			S
Industrial hydraulic fluid	X	S	S	X	X	X	X		S
Metalworking fluids	X			X	X	X	X		X
Turbine oils	X			X	X				
Greases	X			X		X			S

Note: X = Essentially always present in that type of lubricant, S = Sometimes used in that lubricant.

NLGI grade	Dart penetration (0.1 mm)	Consistency and typical applications
000	445–475	These are very soft greases, similar
00	400–430	to thick gear oils. Most often used
0	355–385	in slow gear drives.
1	310–340	These are the types usually used in grease guns. #1 and #2 greases are
2	265–295	typically used in ball and roller brgs. #3 greases are sometimes used in
3	220–250	high temperature ball bearings and in plain bearings.
4	175–205	"Block greases" - very hard, almost
5	130–160	like bricks. Used for sliding applications
6	85–115	and at higher temperatures.

FIGURE 8.10 The NLGI grading system for greases controls the grease consistency and pumpability.

GREASES

Lubricating oils are a combination of the base oil and the additives, whereas greases add a thickener that serves as an oil reservoir. In hydrodynamic and mixed film applications, the thickener acts to slowly let the oil and additives wick out and provide the lubricant film. In boundary lubrication applications, the thickener joins with the oil and additives to provide the separating film.

In manufacturing the grease:

1. Additives typically range up to about 15% of the volume.
2. Thickeners can run from a low of 4% to a maximum of about 20% of the volume.
3. The remaining fraction is the base oil.

There are several ways of categorizing greases, with the most common probably being the National Lubricating Grease Institute's (NLGI) numerical system for grease stiffness shown in Figure 8.10.

This system refers to the pumpability or flowability of the grease and gives little information about its lubricating ability. For that we have to rely on data from the manufacturers, and there is little in the way of publicly recognized standards that quantify the overall applicability of a grease for an application as coupling grease, ball bearing grease, or chassis grease. (There are a tremendous number of standards, such as ASTM D-1742 Oil Separation from Lubricating Grease during Storage, that apply to one specific part of a grease's function, but to the best of my knowledge, there are no comparative public standards for the applicability of greases.)

When a grease is used for ball and roller bearing lubrication, two points to keep in mind are that:

1. The viscosity of the base oil is critical to providing the supporting film in the Hertzian contact region.
2. Most of the data we have seen has shown that the addition of EP additives generally reduces the life of the rolling elements bearings by about 30%.

LUBRICANT FUNCTIONS

ROLLING ELEMENT BEARINGS

Early in this chapter, there was a section explaining how lubricants function in the contact area between the rolling element and the race. In this section, we'll expand a little on that and add the action of the lubricant in the rest of the bearing.

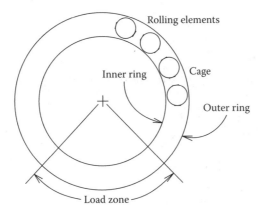

FIGURE 8.11 A cross section view of a typical rollling element bearing with a radial load.

Figure 8.11 shows a cross-sectional view of a rolling element bearing. In this example, the bearing is under a pure radial load and as such:

1. When the rolling elements are in the load zone, they are trapped between the two rings, and the rolling elements drive the cage.
2. As soon as the rolling elements leave the load zone, the cage drives them.

This accounts for one of the lubricant requirements, and that is the need to provide cushioning between the elements and the cage as they go in and out of the load zone.

A second lubricant requirement comes into play when the rolling element returns to the load zone. There is always some minute misalignment between the two rings, and there is always some friction between the rolling elements and the cage in the unloaded portion of the bearing. As a result, there is an almost instantaneous acceleration of the element as it returns to the load zone and, if the lubricant film isn't adequate, skidding occurs, similar to an airplane tire when it first contacts the runway on landing. With time, this skidding roughens the rolling path, and premature failure results.

(This description only applies to bearings that have no or very little thrust load. One of the advantages of the thrust-loaded bearing is that the rollers do not experience this acceleration and deceleration and the resultant skidding. One sure way to tell if these actions have affected the life of a bearing is to look at the load zones after removing the bearing from operation.)

The last need for lubrication is shown in Figure 8.12, in which a heavily loaded and elastically deformed ball is shown in contact with a race that was perfectly machined to contact a lightly loaded ball. The no-longer-round ball has several areas of rolling and skidding, and a separating lubrication film is required to prevent damage.

The Pressure–Viscosity Coefficient

Much of what we see in .ball and roller bearing lubrication also applies to gear lubrication, with the additional confusion that there is also a substantial amount of sliding action. It seems that many lightly loaded bearings and gears could probably be well lubricated by clean corn oil but for demanding, heavily loaded, or elevated-temperature conditions, an understanding of the pressure–viscosity (PV) coefficient is very helpful.

As mentioned in conjunction with the Hertzian contact area, when fluids are exposed to high pressures, their viscosity increases. Figure 8.13 is a summary of how the viscosity varies when a plain mineral oil is subjected to these pressures. We included the honey and antifreeze examples from the earlier chart and, when you consider that a heavily loaded rolling element bearing may

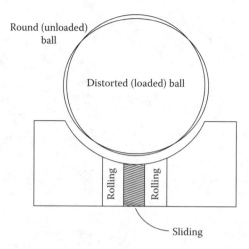

FIGURE 8.12 The defermation of a heavily loaded bearing ball and the effect on the race contact.

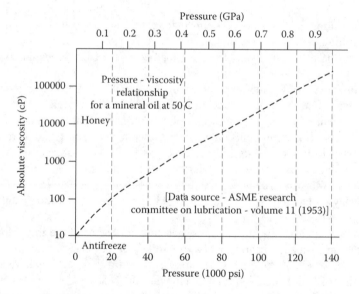

FIGURE 8.13 This chart shows how a 10 wt oil is only slightly more viscous than antifreeze at atmospheric pressure but more viscous than honey at 100,000 psi (0.7 GPa).

see pressures of over 2 GPA (≈ 300,000 psi), the change in viscosity is tremendous. At room temperature, this oil goes from being slightly more viscous than antifreeze to the point where, at less than half the peak pressures seen in some bearings, it is already more than 100 times as viscous as honey.

The PV coefficient is a measure of the relative change in viscosity of lubricants as the pressure increases, and it largely depends on the base oil used in the lubricant. The importance of this is that some base oils have substantially greater PVs and as a result have thicker viscosity conversion films at the operating temperatures and pressures. Table 8.3 gives a selection of lubricating oils and their relative lubricant film thickness at both 25°C (76°F) and 80°C (176°F) at 2,000 bar (≈ 29,000 psi). But there have to be two cautions about using this table:

TABLE 8.3

Oil Type	Viscosity at 2000 bar/Atmospheric Viscosity at 25°C (76°F)	Viscosity at 2000 bar/Atmospheric Viscosity at 80°C (176°F)
Paraffinic mineral	15–100	10–30
Naphthenic mineral	150–800	40–70
Aromatic solvent extracts	1,000–20,000	100–1,000
PAOs	10–50	8–20
Polyether oils	9–3-	7–13
Ester oils	20–50	12–20

Source: Data based on information from Klamann, D., *Lubricants and Related Products,* Verlag Chemie, Weinheim, FRG, 1984.

1. The range of pressures in the chart does not go very high, i.e., 2000 bar when bearings and gears sometimes operate at seven to ten times that pressure.
2. The PV coefficient refers to the increase in viscosity of that lubricant at that temperature and pressure. If the application has oil with a low VI and it is running at the same conditions as the high-VI oil, the latter starts with a big advantage.

(In addition to the chart, we have seen data that indicates that the PV coefficient for PAOs and PAGs [polyalkylene glycols] rises above that for conventional mineral oils above about 75°C [168°F]. This may explain why we have seen improved performance from PAO oils in several high-temperature, heavily loaded roller bearing and gear applications.)

LUBRICANT FILMS AND THE EFFECTS OF CONTAMINATION

There have been many studies that have shown that water contamination reduces the fatigue life of rolling element bearings. From 1960 to date, we are aware of at least seven extensive analyses, none of which had the operating conditions duplicating other studies. Based on them, we feel that there can be a consistent reduction in life at water concentrations as low as 0.01% (100 ppm) and recommend that water levels be kept below half of that (50 ppm).

(One of the problems that water presents is that, as a polar molecule, when it is put under pressure it does not increase in viscosity like lubricant oils. When it contaminates the typical hydrocarbon lubricant and is put under pressure, it tends to vaporize, removing the cushioning layer in the Hertzian contact zone. A second problem is that when the bearing is shut down, the water leads to corrosion.)

Every text on bearings has charts such as that in Figure 8.14, showing that the film thickness increases with bearing life. From a practical standpoint, probably the most important thing that can be done is to pay attention to the lubricant temperatures and the minimum viscosity recommendations in Figure 8.15.

Figure 8.15 takes Figure 8.8 and adds to it the minimum suggested viscosities for several common bearing styles. One important point about this chart is that at low speeds there isn't enough velocity to cause viscosity conversion, and those bearings have to be treated as though they involved sliding contact. Some other general notes are:

1. Roller bearings such as cylindrical, spherical, and tapered rollers require a greater film thickness than ball bearings because they are more sensitive to surface variations, i.e.,

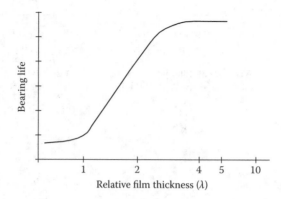

FIGURE 8.14 As the film thickness increases, rolling element bearing life increases dramatically then levels off.

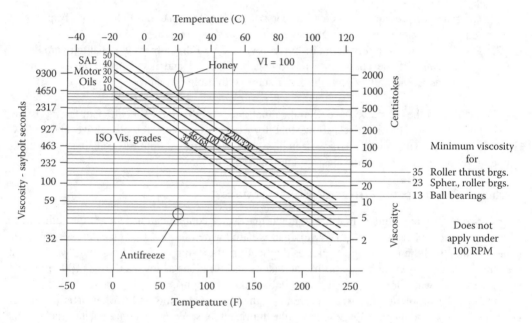

FIGURE 8.15 Expanding on Figure 8.8, this shows the minimum required viscosity for ball, roller, and thrust bearings.

    instead of a point contact, they have line contact, and any disturbance along that line, such as a machining variation or a speck of dirt, reduces bearing life.

2. Roller thrust bearings and spherical roller thrust bearings have radial contact lines that are sensitive to lubricant film thicknesses because the outer edge of the roller is moving much more per revolution than the inner edge. This difference in velocities requires a heavy lubricant film.

PLAIN BEARINGS

In the first part of this chapter, the description of plain bearings indicated that the maximum pressures were in the range of 15 MPa (2200 psi), and the typical operating clearance was in the area of 0.03 mm (0.001 in.). These bearings are very tolerant of small particle and water contamination but very sensitive to misalignment.

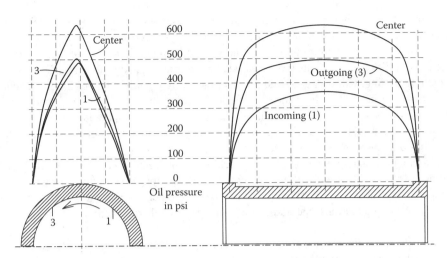

From Beauchamp Tower's original experiment showing oil pressure developed in a
4" diameter × 6" long journal bearing (in psi)
(Kingsbury's experiments on similar devices showed the friction coefficient to be 0.0005)

FIGURE 8.16

Figure 8.16 shows the results of Tower's series of landmark experiments used to determine the pressure profile of a plain bearing. As can be seen in the view on the left, the pressure builds as a result of the rotation of the shaft in the bearing housing, reaching a peak slightly after the center of the bearing. The right diagram shows a side view of the same bearing along with the three profiles of the pressures across it.

From these drawings, it is easy to see the disastrous effects of not only misalignment but also of putting an oil groove or other features that may allow lubricant to escape from the loaded field.

SUMMARY

Lubrication appears fairly simple, but the details are extremely complicated. In this chapter, we have tried to explain some of the more important points of fundamental lubrication mechanisms with emphasis on ball and roller bearings. Given in the following table are a series of general guidelines that can be used for most applications.

Minimum Viscosity at Operating Temperature (cSt)	Equipment
13	Ball bearings and hydraulic systems
23	Spherical, tapered, cylindrical roller bearings, and lightly loaded gears
35	Roller thrust bearings
40	General spur and helical gears
70	Worm reducers

DEFINING WEAR

Wear is the undesired removal of material by contact with another material. There are three different basic wear mechanisms and, from what we have seen, they typically account for between 3% and

7% of the total maintenance cost in the average plant. However, in some plants, such as difficult mining operations, they may amount to almost half of the total cost.

There is disagreement within the failure analysis community as to the wear categories. Some say there are five different mechanisms — surface fatigue, fretting, adhesion, abrasion, and erosion. Others include fretting with corrosion and lump surface fatigue with the other fatigue mechanisms. We have chosen to include both surface fatigue and fretting in our general discussion of fatigue, with additional details on surface fatigue in the section on bearings and additional information on fretting in both the section on bearings and the section on corrosion. Therefore, this section will address adhesion, abrasion, and erosion.

From a practical standpoint, one of the difficulties with wear is that it is difficult to diagnose the specifics of the problem without a metallurgical examination. It certainly isn't hard to tell that something has worn away but identification of precisely whether it is adhesion, abrasion, or erosion isn't easy. Two other situations that add to the complexity of the problem are:

1. Frequently, a situation will involve synergistic mechanisms, i.e., more than one wear mechanism is responsible for the loss of material.
2. It is not uncommon that wear and corrosion mechanisms coexist and act together to accelerate equipment deterioration.

ADHESIVE WEAR

Adhesive wear occurs by transferring material from one surface to another by solid-phase welding that results from the relative motion. When a stainless steel bolt seizes during assembly or a gear such as that shown in Photo 8.1 has radial drag lines, adhesive wear has been there and the pieces have literally welded themselves together at close to ambient temperatures.

In the dedendum of the tooth shown in the photo, the opposing tooth contact has been severe enough that small areas have been welded, then torn out. Other common terms used to describe adhesive wear are *scoring*, *galling*, *scuffing*, and *seizing*.

Detection — The heat of welding almost always causes changes to the material's microstructure that can be seen by a metallurgical examination.

Prevention usually involves using materials that are insoluble in each other, i.e., cannot weld to each other, such as cast iron and hardened steel, or the use of special surface coatings or EP lubricants. Other alternatives involve reducing the local temperature to reduce the possibility of welding and using materials with very smooth surfaces. (This latter approach may sound impractical to some but it is actually one of the methods used to increase the power throughput in precision gear sets.)

PHOTO 8.1 Adhesive wear of a through-hardened gear tooth.

PHOTO 8.2 The bolt has welded itself to the steel flange.

Common practice is, when gears or similar pieces of equipment are meshing against each other and the materials are similar, i.e., they are both steel, the hardness of the pieces should be separated by at least 10 Rockwell points. Photo 8.2 shows a section of a broken bolt that has welded itself to the flange, and it initially looks like an exception to this rule. In this example, the bolt and the flange were separated by 25 HRC points, but three things happened to allow them to weld together:

1. The pieces were at about 200°C (395°F), so they started at temperatures well above ambient.
2. They were not in intermittent contact like the gear teeth. (The advantage of the intermittent contact is that it allows the tooth surfaces to cool before another contact is made.)
3. They had very little lubricant on them that could prevent the welding.

ABRASIVE WEAR

Abrasive wear involves three categories that differ slightly in their mechanisms, but all include removal of material from a solid surface by particles sliding along the surface.

Other terms used to describe abrasive wear include *scouring* and *erosion*, and the three categories are gouging, high-stress grinding, and low-stress scratching.

Gouging results when large particles are removed from the surface, leaving grooves or pits. A good example is the teeth of a power shovel in a quarry that handles large heavy rocks. This requires materials that are relatively hard and also have good fracture toughness, such as Hadfield steel or tempered medium carbon alloy steels. To the best of our knowledge, there is no accurate laboratory test that would give close correlation with field practices.

High-stress grinding (HSG) occurs when there is either plastic flow of ductile or fracture of brittle target materials when the load is high enough to cause fracture of the abrasive medium. The worn surfaces show scratch marks where the material was removed. This is typical of industrial grinding of minerals in ball and rod mills, with fine particles under heavy pressure. Impact resistance is not needed but the target materials need some fracture toughness, and hardened white irons and high-carbon steels can be economically used in most applications. As a laboratory test, the wet-sand erosion test seems to be the most applicable.

Low-stress scratching (LSG) happens when there is no impact loading, and the unit pressures are relatively low. It is what would be seen with a plow in sandy soils where the repeated contact with soil particles would slowly remove target material. In these conditions, the matrix is not critical, as toughness is not important, and materials with hard carbides are generally used. As a laboratory test, the dry-sand erosion test is a good source of comparative data.

Detection of abrasive wear in metals often requires the use of metallography to look at the outermost layer of the grain structure to see if it is deformed from the severe contact stress. Other times, such as in Photo 8.3, it is relatively obvious that the piece has been abraded away. This photo shows a tremendously worn cast iron chain sprocket from a wood yard.

In nonmetallic materials, abrasive wear usually shows grooves or gouges, frequently with small scrapings (or shards) of parent material alongside the damage path.

PHOTO 8.3 Abrasive wear of a large engineered chain sprocket.

Abrasion prevention can involve several approaches:

1. In LSG, the hardness of the target material is not as important as the structure. For example, two materials, hardened steel and hardened cast iron, may have identical HRC 58 hardness readings but the cast iron with its extremely hard carbide "islands" will far outlast the hardened steel. (In the cast iron, the carbides are much harder than the HRC 58 overall hardness reading, and they act to shield the surrounding matrix from the action of the wear particles. The steel is essentially homogeneous and doesn't have the same ability.)
2. In both gouging and HSG, harder target materials with improved fracture toughness are frequently used. Hardfacing (welding an extremely hard and abrasion-resistant material on the surface of a target area) is frequently used in the mining and materials-processing industries to provide local protection, as shown in Photo 8.4. This is a section of a dragline bucket, and the weld overlays are laid on top of the bucket steel in a pattern to reduce the abrasive wear. These overlays are typically very brittle but are supported by the tougher and more ductile parent material.
3. Reduce the pressure and velocity of the abrading particles.

Textbooks frequently mention that changing the abrading particle shape can improve a wear situation but this is usually impossible to do in an operating plant. However, one good lesson on the importance of particle shape in an abrasive or erosive application occurred when we replaced some steel pipes with HDPE piping. These pipes carried sand from our crusher to the settling pond, and the steel pipes repeatedly wore out after only 2 years. Another division of our company made the resins for HDPE pipe, and their data showed that HDPE would outlast steel about 5 to 1 in LSG from sand in water, so we installed several sections of 12-in. HDPE. About 2 months after the installation, we decided to use ultrasonic thickness testing to check on the wear rate of the pipe, and as soon as we looked at the thickness data, we immediately told them to change the pipe. The sand particles were the same size but the wear tests had been conducted with river sand, which is older and smoother compared to the particles that had just been fractured in the crusher, and those sharp edges had attacked the new pipe with a vengeance. When they removed the pipe after operating for only 3 months, there were areas that were down to less than 10% of the original thickness. If anybody asks you, our "research" shows steel outlasts HDPE by about 6 to 1 when used on new sand!

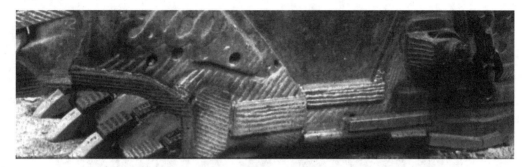

PHOTO 8.4

EROSION

Erosion generally involves flowing loose particles attacking a surface but it is sometimes identified by local terms. The three general categories are:

1. *Airborne erosion* from solid particles carried by the gas flowstream. The particles usually range in size from 0.005 mm to 0.5 mm (0.0002 to 0.02 in.), and the typical airstream's velocity is over 5 m/sec. (15 ft/sec.).
2. *Liquid erosion* from solid particles carried by the liquid flowstream. The solid particles are usually over 0.002 in. diameter with flowstream velocities over 2 m/sec.) (6 ft/sec.).
3. *Liquid erosion* from liquid droplets over 0.25 mm (0.010 in.) carried by a gas stream with velocities dependent on flowstream components.

Erosion is usually identified by a combination of the distinctive turbulent flow patterns left on the attacked material and the knowledge of the impinging stream. Photo 8.5 shows a diaphragm out of a large centrifugal compressor with substantial erosion damage from a wet flowstream at 5000 r/min on a 45-cm (18 in.) radius. Photo 8.6 shows the surface of a pump impeller that suffered erosion from solid particles in the flowstream.

Two of the interesting properties of airborne erosion are that not only do the target and impinging stream angles have an immense effect on the relative wear rates as shown in Figure 8.17, but also the particle velocities have an equally impressive effect. From this, we can look at the relative flow rates and predict almost exactly what effect a change in process flow will have on erosive wear rates.

PHOTO 8.5

PHOTO 8.6 Pump impeller damaged by solid particles in a liquid flowstream.

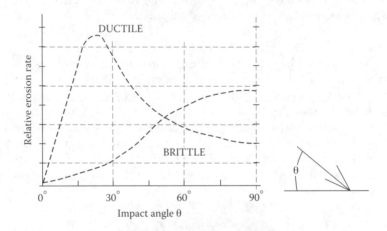

FIGURE 8.17 Relative erosion rate vs. impact angle (from numerous sources based on the work by J.G.A. Bittter).

SUMMARY COMMENTS ON WEAR FAILURES

It is difficult to understand exactly what is happening with a wear failure without a metallographic examination. We've seen numerous examples of changes in materials that haven't solved the problem and have sometimes made the situation worse. However, with abrasion and airborne and liquid erosion, changes in operating conditions will have very predictable effects on component life.

9 Belt Drives

Belts were developed to provide a means for transmitting power from the prime mover — a steam engine or a water wheel — to the point where it was actually needed. The first belts were used in water-powered mills, and they were generally flat, made from leather, and ran on wooden pulleys that had a very slight crown to keep the belt in position. In those early days, a mill would have a line shaft driven by the prime mover. Then, off the main line shaft, there would be many flat belts, each supplying a smaller line shaft, a machine, or a part of the process. One of the truly impressive sights of my early days in industry was watching the "head operator" using a special stick-like tool to put a 600 mm wide, 9 m long (24-in.-wide, 30-ft-long) flat belt on a line shaft pulley rotating at 500 r/min. A huge motor drove the main line shaft, which in turn drove 15 other machines.

Before the modern variable-speed motors were introduced, AC motors only ran at given fractions of their synchronous speed (about 3600, 1800, 1200, 900 r/min, etc., in the U.S., and 3000, 1500, 1000, 750 r/min in Europe and about those areas with 50 cycle power). If the machine needed an operating speed other than one of these, some method of changing the speed had to be made. This could be done with gears or chains but, compared to V-belts, they are expensive, noisy, complicated, and require lubrication. (However, both offer precise speed regulation, something that is impossible with V-belts.)

V-belts were first used commercially shortly before 1920. In the U.S., Gates made the first industrial V-belts, and Dayco (now Carlisle) pioneered the automotive V-belt. The first synchronous belts were developed in the 1940s to ensure proper timing of the needles and bobbins in industrial sewing machines. In the last 40 or so years, there have been tremendous changes in belt technology with the "modern V-belts" showing up in the 1960s, raw-edged notched V-belts and the first serpentine belts showing up in the 1970s, and many variations on the original square-toothed synchronous belts.

BELT DESIGN

Figure 9.1 shows the major V-belt components. The original construction used a tension member made from cotton or linen, the elastomer was natural rubber, and the canvas wrapping was really needed to help keep the parts from separating. Through the 1950s and 1960s, the tension members were generally made from rayon. Today, tension members are almost exclusively made from polyester, though many alternatives are available for special purposes. The *elastomer* is a synthetic rubber compound (usually either SBR or neoprene), and wrappers are becoming less common and generally used in specific applications.

The classical V-belts were developed years ago. The modern 3V, 5V, and 8V belts that were introduced in the 1960s are narrower and deeper than the classical belts. Figure 9.2 shows how the different belts compare in their relative sizes and load capacity.

In addition to the change in shape, another change in the construction of V-belts involved the introduction of "molded notch" belts that Dayco introduced in the 1970s. These have the same basic dimensions and act in the same manner as modern and classical V-belts, except that some of the material in the bottom of the belt has been notched out, allowing the belts to wrap around smaller sheaves. Although it is true that these belts can be used on smaller-diameter sheaves, they

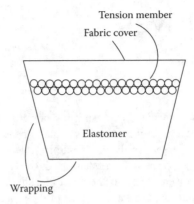

FIGURE 9.1 Typical V-belt construction.

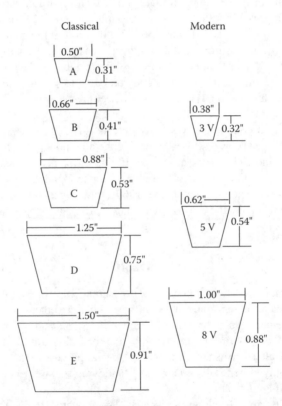

FIGURE 9.2 Comparing classical and modern V-belts shown from top to bottom in relation to their load carrying ability.

also run cooler on standard-size sheaves and take less energy to flex. (There is a group of power transmission [PT] belts used on variable-speed machinery, snowmobiles, and some ATVs that looks a little like a molded notch belt. However, putting them side by side, it can be seen that the PT belt is wider, thicker, and much more rigid than a V-belt.)

In addition to the obvious difference between raw-edged and wrapped V-belts, there are internal differences.

PHOTO 9.1 Two sets of banded V-belts with classical belts on the left and raw edged, notched belts on the right.

- Most wrapped belts use SBR as their elastomer but raw-edged belts use neoprene, which has better heat resistance and stability.
- Raw-edged belts have more tension members, are generally stronger, more temperature resistant, and more efficient, but they cost about 25% more.
- Because of the difference in friction coefficient, raw-edged belts are much more sensitive to badly worn sheaves and will slip more.

Photo 9.1 shows a pair of "banded V-belt" sets, and the differences in basic construction and size are obvious. The set on the left is a pair of conventional wrapped B belts, and the set on the right is a pair of raw-edged 3V belts. Although there is a huge difference in size, the two sets have approximately the same capacity.

There are many other belt shapes, such as double-sided V, V-ribbed (serpentine), metric, etc. They all have different applications, and they can be analyzed with many of the same approaches used on V-belts.

Some of the internal differences in synchronous belts are:

- Conventional synchronous belts, with square or rectangular teeth, have fiberglass tension members. The high-performance types, with their rounded teeth, use Kevlar™ or other aramid fibers.
- Kevlar™ belts require more initial tension than fiberglass belts because there is some initial fiber stretch in the Kevlar™.
- Both types of synchronous belts use neoprene as the elastomer.

Belt Drive Design and Life

Sizing a V-belt is not terribly complicated, and most manufacturers have not only catalogs with the design steps clearly outlined but also computerized design programs that are readily available. The major steps in designing a V-belt drive that should last 25,000 h are:

1. Start with the motor (prime mover) horsepower and r/min. Based on that, select the basic belt size.
2. Look up a service factor (SF_m) from a chart that lists the prime mover design, the type of application, and the number of hours per day. (An easy-starting smooth load such as a fan would have a lower SF_m than a log chipper that might start up with a heavy load and may see big load variations during operation. In addition, a unit that only runs occasionally would have a lower SF_m than one that is expected to run 24 h per day.) Multiply the horsepower from Step 2 by SF_m.

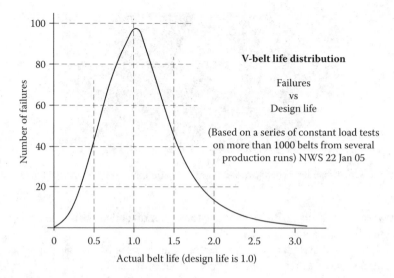

FIGURE 9.3 V-belt life distribution.

3. Next, select the reduction ratio, and look in the chart for the sheave sizes for that ratio. Then select the belt length and the center-to-center spacing and, from that, find the application factor (AF). The size of the sheave is important because the larger sheave has less force, while the center distance of the sheaves affects the number of fatigue cycles.
4. In determining this application factor, consider the contact angle. This is the distance the belt wraps around the smaller sheave or sprocket. As the wrap falls substantially below 180°, or a given number of teeth for a synchronous belt, the horsepower capacity has to be lowered.
5. Multiply the horsepower from Step 3 (prime mover hp $\times$ SF_m) times the Application Factor (AF) to get the design horsepower. Divide the design horsepower by the horse-power capacity of the belt for the number of belts required.

The design process for a synchronous belt drive is similar except that the drive life will be about 12,000 h. To reach a design life of 25,000 h with a synchronous belt drive, an addition of about 0.4 to the SF_m would be needed.

As shown in Figure 9.3, even though the V-belt design life is 25,000 h, because of manufacturing, operational, installation, and maintenance differences, there are tremendous variations in the lives of individual belts.

> In the early 1970s, when we called the belt manufacturers and asked them how long a properly designed (according to their catalog) belt application would last, they said 25,000 h. By the mid-1980s, when asked the same question, they typically said 15,000 to 20,000 h. In the late 1990s and the early years of the 21st century, the answer to that question was, "It depends." When pressed, they said the life would "generally be in the range of 15,000 to 18,000 h," stating that irregularities in operation frequently prevented them from reaching their 25,000-h goal.

BELT OPERATION

As a V-belt wraps around the sheave, the elastomer bulges outward, transmitting power between the sheave and the belt tension member. Within the belt, the loads should be transmitted evenly from the elastomer to the tension member, which transfers the load to the other sheave, and each section of the belt is stressed hundreds of times a minute. In this continuous process, the fibers in the tension member will eventually fail from fatigue, and a critical point for long tension member life is even distribution of the load across the entire belt width, eliminating excessive stress in any of the individual strands. The angles on the side of the V-belt groove change depending on the size of the sheave. Larger sheaves have straighter sides because the belt is wrapped over a larger radius, and the sides of the belts don't bulge out as much.

Flat belt operation initially appears similar to V-belt operation but, because the flat belt doesn't have the wedging action and relies solely on friction between the belt and sheave surfaces, they typically need much greater tension to prevent slippage. The typical tension ratio for good flat-belt operation is on the order of 2 to 1, and the resultant bearing and shaft loads are much higher than those of the V-belt drive.

Synchronous belts, with their teeth fitting the drive and driven sprockets, generally don't require as much tension as either V-belts or flat belts. Nevertheless, because there is the positive engagement (and the danger of tooth jumping) and because the tension members have some initial stretch, belt vendors generally recommend that the total tension in a synchronous drive be the same as the equivalent V-belt drive. However, unlike V-belts, synchronous belts don't require periodic retensioning.

Figure 9.4 shows the stress cycles that the tension member fibers experience as a belt rotates. Starting on the slack or loose side at A, the belt is deformed as it rotates around the sheave, creating a tensile stress in the tension member fibers. As a point on the belt travels around the driven sheave toward the tight side, the tension remains fairly constant until about the midpoint of the rotation when it starts to increase because of the load that has to be accommodated. When the point reaches B, the full tight-side tension load is realized, but the stress from bending around the sheave disappears. Then the belt travels to C where, not only does it see the full tight-side tension load, but also the bending load from rotating around a sheave that is smaller than the driven sheave. It then rotates with the sheave until it reaches D, at which the bending stress disappears and only the slack-side tension remains. In this travel sequence around the sheaves, the tension member sees two very distinct tension peaks repeated with every cycle, and tension member (belt) life is dependent on both the magnitude of the peak values and the number of fatigue cycles. The basic stress cycles seen by a V-belt and a synchronous belt are essentially identical but, because the synchronous belt is thinner, the stress contribution from bending is much less.

One point that is very obvious from Figure 9.4 is that the use of backside idlers can be extremely damaging from a fatigue standpoint. These idlers cause stress reversals in the tension members, and the result usually is a greatly shortened belt life.

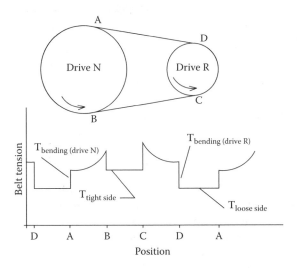

FIGURE 9.4 This figure shows the stress cycles the tension numbers go through during a revolution.

DRIVE EFFICIENCIES

According to several belt manufacturers, some typical belt drive efficiencies are:

- New V-belts, just installed — 97%
- Typical V-belts with periodic retensioning — 93% to 94% (Raw-edged belts are typically about 1% more efficient than wrapped belts.)
- Worn sheaves with loose and flopping V-belts — 88% to 90%
- Synchronous belts — 97%+
- Flat belts — 97% when properly installed but maintaining correct tension is important.

TEMPERATURE EFFECTS

Temperature is one of the keys to long belt life, and the cooler a belt runs, the longer it will last. All of the belt manufacturers have run extensive studies on temperature effects, and Gates' data shows that, beginning at about +35°C (≈95°F), belt life is halved by every 10°C (18°F) increase in belt temperature.

The minimum belt start-up temperature is –35°C (–30°F) and is governed by the flexibility of the tension members. But that only applies to the start-up temperature because once the belt is running it will generate enough heat internally to keep it well above this figure. A good rule of thumb is that the operating belt temperature should be between 33° and 50°C (90° and 120°F) and above 60°C (140°F) or so, very rapid degradation occurs.

Elevated temperature degradation is insidious. The first thing that happens is that the elastomer gets softer. This prevents the load from being transmitted evenly across the tension members to the point where the center members are essentially unloaded. Then, with time at temperature, the elastomer ages and hardens until it becomes bricklike. It can't compensate for the normal wear, needs higher tensions to prevent slippage, and when it cracks, doesn't evenly transmit the loads. Adding to the problem is the fact that, as the belt temperature increases, the fatigue strength of the tension member fibers decreases.

BELT OPERATION AND FAILURE CAUSES

A good understanding of the fatigue stresses shown in Figure 9.4 is an important aid to appreciating how and why temperatures affect power transmission belt failures. Belt manufacturers say that the major cause of failures is elevated temperature. In addition, they indicate that the top five causes of elevated temperatures in V-belts, in order of their importance, are:

1. Slippage
2. High ambient operating temperatures
3. High operating loads
4. Misalignment
5. Belt bending stresses

Slippage

For a V-belt to function there is always some slippage (creep) between the belt and the sheave. As mentioned earlier, as a portion of the belt comes into contact with the sheave, it wedges itself in, rotates a distance with the sheave, and then has to be pulled out. In an ideal application, the creep (slippage) is about 0.5%, and as the slippage becomes greater, more and more heat is generated.

Slippage isn't difficult to control, but it does require attention on a regular basis. Over the first few hours of operation, the belts become seated (wear in) on the sheaves, and there is also some minor initial wear of the sheaves. The most rapid tension member stretch and belt tension relaxation happens in the hour or two after startup, and the rate of stretch rapidly decreases over the first 24 h.

However, not only is the tension member always very slowly stretching, but also the elastomer and the sheave are slowly wearing, so V-belt tension has to be periodically adjusted for good reliability and operating economy. Good practice says the belts should be tensioned at startup, checked about a half hour later, and then retensioned within the first million or so belt cycles. They should also be retensioned periodically whenever the slippage becomes excessive but by the time it is announced by squealing, the belt life has been tremendously reduced. How do you determine when slippage is excessive? Check the sheave speeds with a good quality strobe light, and when the slippage reaches 4% you know that retensioning should be scheduled.

The following chart is a guide written for a belt drive that has been in operation for more than 1 week, i.e., the belts and sheaves have worn in and the initial stretch has occurred. It shows the percent belt slip and relates that to a recommended schedule for retensioning.

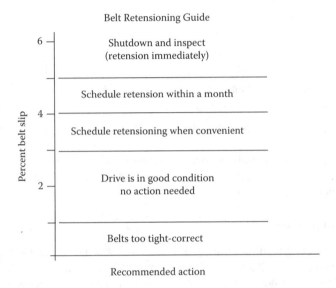

Tension Ratios

Tension ratios are important to V-belt life and, although the term sounds complicated, it is actually fairly simple to understand and to measure. The *tension ratio* is the ratio of the tight-side tension divided by the loose-side tension during operation. When the shafts are at rest, the tensions are equal. But to get the shafts to rotate, the tight-side force has to be more than the loose-side force. Comparing these two forces gives us the tension ratio. If the belt is extremely slack and flopping, the slack-side force (the tension) is very low, and the ratio between that force and the tight-side force is high. The more tension there is on the slack side, the lower the tension ratio, and many industry experts say the ideal would be to maintain the tension ratio between 5 and 9. If the tension ratio goes lower, the force on the bearings increases and the assembly suffers from short belt and bearing life. If the tension ratio is too high, the belts slip and fail. (The ideal flat belt tension ratio is about 2.)

Figure 9.5 shows how critical the tension ratio is in preventing belt slippage. It shows that, as the ratio gets larger, the slip increases, with the result that the belt gets hotter and hotter and the life decreases rapidly. For example, if the drive sheave is 4 in. in diameter and rotates at 1000 r/min, theoretically an 8-in.-diameter driven sheave should operate at 500 r/min, but because of creep this doesn't happen. Instead, if one inspects a very-well-installed sheave, one finds that it will run at about 497 r/min but a sloppy, worn, and badly mistensioned application might be turning as slowly as 450 r/min. The easiest way to determine the percent slippage is to compare the sheave speeds with a strobe light or a digital tachometer. (In addition, by using a strobe light and an infrared

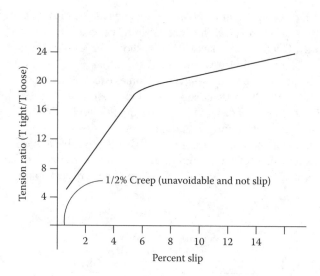

FIGURE 9.5 Tension ratios vs. % slip for a V-belt drive.

thermometer, a belt drive can be inspected almost as effectively as a bearing can be inspected with a vibration analyzer.)

One interesting comparison point in understanding the difference in efficiency between synchronous and V-belts is that it takes energy to generate heat, and a major advantage of synchronous belts is that they don't slip.

High Ambient Temperatures

One of the truly frustrating features of many plant operations is the belt guard that looks as though it was designed to protect Fort Knox. It isn't uncommon to see belt guards in which it is absolutely impossible to see what is inside — but it's obvious that the belt is hot because the guard is warm to the touch. The designers design these belt guards to positively prevent anyone putting their fingers into the sheaves but there is nothing that says that an intelligent and safe belt guard can't be fabricated from perforated stock or expanded metal, allowing both cooling airflow and easy inspection.

In a published Gates' study, they found that a 36°F increase in the ambient temperature around a belt reduced the belt life by half. It seems like a relatively simple area to address, but we frequently see applications in which there is no consideration of ambient temperatures e.g., those vault-like belt guards mentioned earlier.

> In conducting a failure analysis on a belt drive, if the application is in an area that is uncomfortably hot to you, it is also at a temperature level that is reducing the belt life. Take a minute to look at the installation. Could shields be installed? How about cooling vents? Could the shaft have a deflector installed? Should the area be vented? A simple change we have made to several installations is the addition of a fan blade, such as a motor cooling fan, on the shaft between the belt and the source of process heat.

High Operating Loads

The more power a belt drive transmits, the harder the drive works, and the more heat it will generate. If a well-adjusted belt drive is carrying 133 kw (100 hp), it will have losses of about 8 kw (6 hp) and about 5 kw (4 hp) of that is in internal heat from the work, unavoidable slippage, and flexing. On the other hand, if that same drive is transmitting only 67 kw (50 hp), it will generate about 2 kw (1.5 hp) in heat. This means the belts on the smaller driver will be running much cooler (and will last much longer) than its 133 kw (100-hp) twin.

Belt manufacturers no longer publish factors on the specific effects of load on belt life, but one can see from past data that it is an inverse function close to what is used on roller bearings, i.e., double the load and cut the life by a factor of ten. This brings up the question of loose, mismatched, or missing belts.

- If a drive is missing one of several belts, the other belts are more heavily loaded, and their life will be reduced by a calculable factor.
- If the belts in the set are all from the same manufacturing lot, they will work in to fit even though one or two seem to be loose on startup. There are tight dimensional controls on the set's manufacturing and, even though some manufacturer's lengths are not absolutely identical, they are within a narrow range. In operation, the normal stretch of the polyester tension members will eventually distribute the load evenly across the set.

Misalignment

One of the things every maintenance mechanic is told is that sheaves have to be "well aligned" or "aligned within 1.5 mm" (≈1/16 in.) or some such data. The reason why those sheaves have to be aligned is fairly simple. Misalignment creates additional heat in the following steps:

1. The belt rubs against the side of the sheave groove on its way into position, causing friction that increases belt temperature.
2. The tension members in the belt are flexed sideward, something that they are not designed to do, and work and movement generate heat.
3. The misalignment causes vibration that results in additional loads. The harder the belt works, the more heat it generates.

The alignment specifications for good reliability vary with the belt type. For example:

- With individual V-belt drives, the two sheaves should be aligned within 5 mm/m (0.06 in./ft) of center distance.
- For banded V-belts, because of their more rigid tension members, the alignment should be within 3 mm/m (0.040 in./ft) of C–C distance.
- Synchronous belts, because of the need to mesh the belt and sprocket teeth, require alignment within 2.5 mm/m (0.03 in./ft).

Any misalignment more than these means that either shortened belt life has to be accepted or corrective action has to be taken.

In addition, before checking the alignment, the sheaves and sprockets should be inspected for excessive runout. Their maximum allowable runout, both axial and radial, varies with diameter but a good rule of thumb is 0.3 mm (0.012 in.). Again, units with more runout will operate, but they will also result in shortened belt life.

Sheaves and Sprockets — Sizes and Wear

Figure 9.6 gives a clear demonstration of the improvement in life that can result from increasing sheave size. The peak stresses are developed from a chart such as that given in Figure 9.4. Belts are rated based on the tension member fatigue cycles, and reducing the load results in huge increases in that life.

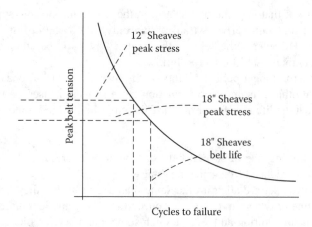

FIGURE 9.6 Effect of belt tension (and fiber stress) on life.

The tension members see stresses from both the force needed to turn the sheave and the forces wrapping the belts around the sheave. For example, if the belts used in the drive are transmitting 67 kw (50 hp) at 1780 r/min:

- The force needed to rotate the 300 mm (12-in.) sheave is 1312 N (295 lb) and, with a tension ratio of 6, this means the tight-side tension is 1574 N (354 lb) and the loose-side tension is 262 N (59 lb).
- Changing those sheaves to 450 mm (18 in.), the force needed to rotate the sheaves is reduced to 876 N (197 lb) and, with the same tension ratio, the tight-side tension is 1050 N (236 lb) and the loose-side-tension is 174 N (39 lb).

With the change in diameter, and the 33% reduction in total tension, the life of the belt set is increased by more than a factor of four. In addition, the belts don't have to be adjusted as frequently, and bearing life is also increased.

Manufacturer's data states that worn sheaves can reduce belt life by 50%. However, that only tells part of the story. As the belt wraps around the sheave and the sides bulge outward to grip the sheave, if the sheave is worn the load needed to provide enough friction to prevent slippage has to be increased. This additional tension load reduces the belt life. The increased resultant force is then applied to the bearings, and their lives are also exponentially reduced. The third negative point about worn sheaves is that they can contribute greatly to a system in which the power transmission losses can exceed 10%. Sheaves usually last through three to five belt changes before they have to be changed, but to get the best results, we should do the following:

1. Use a sheave gauge and, on conventional V-belts, be sure the wear is less than 0.75 mm (0.030 in.).
2. If the application uses banded belts, the sheave wear should be limited to less than 0.5 mm (0.020 in.) because more severe wear will cut the belts loose from the band (and you won't have a banded belt anymore).
3. Synchronous belt sprockets should have no readily visible wear on the teeth.

Belt Resonance

This has become very common with variable-speed motors, and most belt manufacturers classify it as their biggest field problem. What happens is that, as the speeds, loads, and temperatures vary,

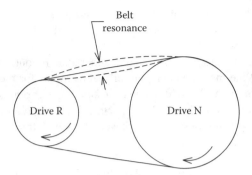

FIGURE 9.7 The belt resonance shown above is almost unavoidable on a variable speed drive.

the resonant frequency of the belts also varies, and it is essentially impossible to design a drive system that doesn't occasionally have the belt resonance seen in Figure 9.7.

Resonance can affect any belt drive, and a simple method of preventing it involves adding a damping device, i.e., a skid plate that just misses the belts during normal operation or another sheave, a kiss idler that just barely touches them. Then, when the belts reach a resonant frequency, the damping device/idler will prevent the vibration from becoming excessive. However, if the skid plate or sheave is installed exactly on the centerline, there is the possibility of two harmonics. Better to install it about 40 to 45% across the centerline.

BELT DRIVE FAILURE ANALYSIS PROCEDURES

The following chart shows the most common causes of belt failures, and following it are a diagnostic outline and a series of belt photos showing and explaining some of the common failure mechanisms.

Cause Symptom	Sheaves or Sprockets Misaligned	Belts Loose	Sheaves Worn or Damaged	Design Weakness	Overloaded	Wrong Belt Size	Pulsating Drive hp	High Op Temperature	Belt Contaminated	Resonance	Installation Damage	Rubbing
Belt slippage		XX	XX		X	X			X			
Cover ruptured											XX	X
Belt turned or jumping	X	X		Poor support			XX			XX		
Elastomer cracked		X		Sheave diam				XX	X			
Bottomed in sheave			XX			X						
Tension member separation	XX		X								XX	
Synchronous belt tooth wear	XX				X		X			X		

Note: XX indicates most likely causes (Don't forget to look for multiple causes.); X indicates another possible cause.

Analyzing a failed belt drive (It would be ideal if there was the ability to have known the belt temperature during operation, but we are going to assume that is impossible.):

1. Rotate the sheaves to measure their axial and radial runout, then check the sheave alignment. Are all the measurements within specifications for the belt type?
2. Inspect the sheave sidewall condition. (Is it smooth and polished or has there been damage, leaving it rough and irregular?) Then, using a sheave gage on V-belt drives or a micrometer on synchronous drives, measure to see if the sheaves/sprockets are in acceptable condition.
3. Examine the broken V-belt.
   a. What do the ends of the tension members look like? Are they unfrayed or only slightly frayed (indicative of a very-short-term overload failure as shown in Photo 9.2) or are they severely frayed (indicative of a long-term failure as shown in Photo 9.4)?
   b. Examine the elastomer.
      i.   Are there cracks in it, i.e., showing the result of slippage and/or high operating temperatures and degraded elastomer?
      ii. Does the elastomer feel soft or greasy (indicative of foreign material contamination from grease, oil, chemicals, belt dressing, etc.)?
      iii. Are there some random areas that appear charred and worn? (Suspect spin burns [see Photo 9.9], i.e., places where the belt stopped for a short time while the sheaves continued turning — with the result that the belt became overheated.)
   c. Do the sidewalls show polishing (indicative of slippage)?
4. Is the cover separated (indicative of poor installation practices) as shown in Photo 9.3?

PHOTO 9.2 The instantaneous failure of a group of four belts. The fact that there is absolutely no fraying of the tension members tells us that the failure happened very rapidly. This is similar to an overload failure of a metal piece and tells us that the physical cause of the failure was present immediately before the failure occurred.

PHOTO 9.3 The cover separation on the right side of the sheave is obvious. This is the result of improper installation procedures.

PHOTO 9.4 The separated cover leads us to suspect an installation problem although it may have been a manufactuing defect. The badly frayed tension members indicate that the failure didn't happen instantly and that the belt has been running with the defect. This is similar to a fatigue failure in a metal part.

PHOTO 9.5 This is a pile of belts that has just been removed from a large fan. The elastomer on the sheave contact surfaces is shiny and polished, proof that the belts have been slipping.

PHOTO 9.6 The underlying construction fabric on the right side of these synchronous belt teeth is showing, and the uneven wear is obvious evidence of misalignment.

PHOTO 9.7 The worn teeth on this synchronous belt are coming off, proof that the belt has reached the end of its design life. The evenness of the fractures shows that the alignment was excellent.

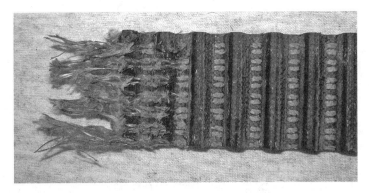

PHOTO 9.8 The heavy wear on the synchronous belt should have been readily visible during a strobe light inspection anytime in the past month or two. The belt eventually failed as the badly worn teeth were ripped off.

PHOTO 9.9 Spin burn damage on an A section belt. In this case, the machine jammed and the belt tension was so loose that the drive sheave slipped and heated the belt literally until it charred. This damage is common when a machine jams but it can also happen from lack of tensioning. During operation, a belt constantly stretches and, on a machine such as a fan, when the tension ratio goes over about 16 (5% slip), the chance of this happening increases rapidly.

10 Ball and Roller Bearings

The first documented use of roller bearings that we could find was in 300 BC when the Greek engineer Diades developed a roller-supported battering ram. Another early application can be seen in the Danish National Museum, where there is a series of rollers that are over 2000 years old.

The modern (and reliable) life of rolling element bearings really began in the late 1800s, but there are examples of common usage from the 1700s. In the 1750s, Dutch windmill center post bearings were using cast-iron balls, whereas the latter part of the 18th century saw some of the better carriage axles using loose cast bearing balls. But the problem with these applications was the lack of ball precision and uniformity. The balls were not exactly the same diameter and the typical difference in diameters was greater than the rolling element deformation under load. This caused heavy individual element loading with the resultant rapid fatigue, spalling, and failure.

Then in 1886, in Germany, Friedrich Fischer designed and built the first plant for the production of precision bearing balls. Additional important early developments included Henry Timken's invention of the tapered roller bearing in Missouri in 1898, and Sven Wingquist's invention of the first self-aligning ball bearings in 1907 in Sweden.

BEARING MATERIALS

To withstand the extremely high Hertzian fatigue stresses and give relatively long life, the steel used in rolling element bearings has to be very hard. There is some alloy variation, i.e., the chemistry of the steel changes a little depending on size of the bearing because the size of a material affects the heat-treating response. However, most bearings are through hardened and are made from either SAE 52100 (Japan's SUJ2 steel) or very similar alloys. There are also surface-hardened (case-hardened) bearings that are most commonly made from either AISI grade 4118H or 4320H, or their European and Japanese counterparts. In addition, there are some relatively uncommon bearing steels that have other additions to improve their toughness and fatigue resistance, and there are also a number of ceramics that are being used for rolling elements and as coatings.

The alloy composition of the through-hardened steels varies a little in the amount of silicon and manganese, whereas surface-hardened ones vary in the amount of nickel and chrome. These variations are relatively small and are not critical to the user, but a good understanding of the basic differences between the two structures is very important, and Figure 10.1 shows this.

Surface-hardened bearings are a little more costly, but in most applications there is no perceptible performance difference between through- and surface-hardened bearings. However, in applications in which the bearing experiences considerable shock loading, the surface-hardened units have an advantage because cracks do not readily propagate through the softer core and across the entire race. Photo 10.1 and Photo 10.2 compare a case-hardened tapered roller bearing with a through-hardened spherical roller bearing. Both have cracked, but the two-year-old tapered roller bearing is still in usable condition because the crack has not propagated through the core, whereas the two-week-old spherical bearing is only a short time from a catastrophic failure. Common situations in which case-hardened bearings are used include paper machine dryer bearings that are subjected to thermal shocks and logging and mining equipment that experience impact loading.

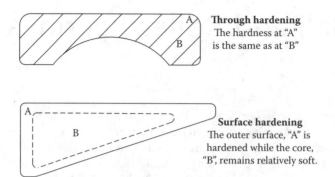

Through hardening
The hardness at "A"
is the same as at "B"

Surface hardening
The outer surface, "A" is
hardened while the core,
"B", remains relatively soft.

FIGURE 10.1 Cross section views of through and surface-hardened bearing rings.

PHOTO 10.1 Note the crack in this surface-hardened race. It has not propagated through to the roller path.

PHOTO 10.2 This shows the shaft fit from a through-hardened bearing. The crack extends to the rolling surface and a catastrophic failure is not far away.

When metals become harder and stronger, their fatigue life increases. Although the rolling elements may be a bit softer than the races, the surface hardness of these bearing components ranges between HRC 58 and 62. Reducing this hardness reduces the fatigue strength of the bearing and a long-used rule of thumb is that for every point the hardness is reduced below HRC 58, the bearing life is cut by about 10%.

> Why did Harry Timken start making his tapered roller bearings case-hardened instead of through-hardened? We've talked with a lot of people and asked them the same question without a definitive answer. However, the consensus is that the common way of hardening components in Henry Timken's world was pack carburizing, and that was the process he used; whereas in Germany, it was common to through-harden materials.

The normal maximum operating temperature at which conventional bearing steels can be used for long periods of time without a reduction in capacity or reliability is about 120°C (250°F). (The steel doesn't actually begin to weaken for about another 50°C, but the bearing operating stresses will create additional heat, and it is wise to leave a reasonable safety margin.) The M50 and 440C bearings can be regularly used as high as 220°C (425°F), and lightly loaded applications can run up to 310°C (600°F).

More valuable for most industrial applications, several very effective coatings have been developed to improve life in wet, corrosive, and contaminated operations. Among others, these include Torrington's TDC (thin dense chrome coated) bearings and Timken's AquaSpexx. Our experience with these is that a fourfold increase in life is not unusual.

In addition, there are some specialty bearing materials used for corrosive, high-temperature, and exotic services, but they usually are not seen in general industrial applications. Stainless steels, such as AISI M50 tool steel and AISI Type 440 C stainless steel, polyamide, ceramics, and nylon, are all used for these applications. One increasing use of ceramics such as silicon nitride is the rolling elements of bearings designed for electrical isolation of the inner and outer rings and these are becoming popular in variable speed motor applications.

PARTS OF A BEARING

To ensure we're all thinking along the same lines, Photo 10.3 shows and identifies the parts of a typical ball bearing. One point we should make is that the pieces we call the inner and outer rings, most persons in general industry identify as the inner and outer races. We use the term ring because it agrees with the terms used in the bearing manufacturing industry.

CAGES

The functions of the cage are to maintain the proper spacing between the rolling elements, to minimize friction, and to distribute the rolling elements' load-carrying ability evenly around the

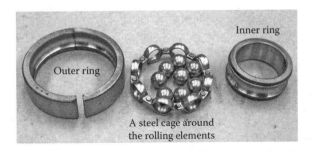

PHOTO 10.3 The components of a ball bearing.

races. There are dozens of cage materials, and the proper selection can be critical to both the bearing's performance and the purchaser's wallet.

For many years, the most common cages were made from stamped steel with the two halves riveted together. (This is called a *Conrad bearing* in honor of the person who developed the machine to assemble the cages.) However, a recent discovery is that cage design is critical because, in many grease-lubricated applications, the cage carries the lubricant that resupplies the rolling elements during operation.

> One of the effects of this discovery is that bearing companies are looking at redesigning cages so they carry additional lubricant and increase bearing life. In turn, this has spurred the design of new cage shapes and the development of polymer cages, and it will be interesting to see what develops.

Nevertheless, the classical advantages and disadvantages of some cage materials are:

- *Stamped Steel* — Low cost, poor frictional characteristics when there is inadequate lubrication, good thermal expansion compatibility with steel bearing components, relatively weak and deformable.
- *Bronze* — Excellent natural lubricity for applications in which lubricant is marginal, expensive, cast structure lends itself to high strength.
- *Polymers* — Light weight, self-lubricating, easily produced with good corrosion resistance, polymer cages are frequently a composite of two or more materials with the properties varying depending on the specific materials used. Common cage materials are nylon, polyethylene, polyamide, etc., with the advantage that they can be formulated to be strong, have good water resistance, high hardness, or high strength at elevated temperatures as needed.

BEARING RATINGS AND EQUIPMENT DESIGN

To determine the life of a ball or roller bearing, the bearing companies initially used a huge sample of identical bearings, all lubricated in the same manner and subjected to the same loads. These bearings were then run at selected speeds with controlled temperatures and ideal lubrication, and detailed records were kept of their failure rates. The combination of the results from many similar experiments enabled the development of the Lundgren–Palmgren equations for rating bearing lives, and, in most cases, the bearing life is specified as a certain rated load for one million revolutions.

The bell curve in Figure 10.2 shows the life distribution that resulted from these experiments. On the curve are marked the L_{10} life, the point at which 10% of the bearings have failed, and the L_{50} life, the average bearing life. It is interesting to note that the curve is distorted a little to the right and there are some bearings that seem to last forever.

From a practical standpoint, some important points to understand are:

1. Most equipment designs are predicated on the L_{10} life of the bearings.
2. The L_{10} life calculated by these experiments and the Lundgren–Palmgren equations is based on failure by Hertzian fatigue.
3. The L_{50} life, the point where 50% have failed, for ball bearings is about 4.5 times the L_{10} life.

Catalog data shows typical L_{10} lives for some common equipment are:

Domestic hand tools: 1500 h
Domestic electric motors: 1500 h

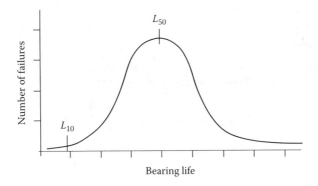

FIGURE 10.2 A "bell curve" showing the typical distribution of bearing failures while operating under identical conditions.

Industrial motors (small): 20,000 h
Industrial motors (large — 50 kw (40 hp) and up): 50,000 h
Industrial centrifugal compressors: 80,000 h
Fans (general industrial): 40,000 h
Mine ventilation: 80,000+ h
Industrial reducers: 5,000 h (but modified by a service factor)
Pickup trucks (light-duty): 250,000 km (150,000 mi)
Pickup trucks (heavy-duty): 360,000 km (220,000 mi)

Using these guidelines, if we were to select a rear wheel bearing for a heavy-duty pickup truck, we would first find the expected loads and from that, go to the bearing selection charts for a bearing that would fit the bearing housing and have an L_{10} life of 360,000 km (220,000 mi).

Understanding the effects of changing loads on a rolling element bearing is absolutely critical to realizing good reliability. For a ball bearing, the bearing life is rated using the formula

$$L_{10} \ (life) = [Basic \ Dynamic \ Load \ Rating/Dynamic \ Equivalent \ Load]^3$$

Therefore, from the formula, we see that doubling the load on a ball bearing will cut the Hertzian fatigue life by a factor of 8. In a similar manner, the exponent used for roller bearing life calculations is 3.3, and doubling the load on a roller bearing will cut the life by a factor of 10.

So if a ball bearing has a basic dynamic rating (BDR) of 454 kg (1000 lb) and the actual load is 100 kg (220 lb), the calculation for the design life would be:

$$L_{10} = [454/100]^3 \times 1,000,000 = 93,600,000 \ cycles$$

and

$$L_{50} = 402,000,000 \ cycles$$

If the load changes, the life will exponentially increase or decrease. For example, if the load on that ball bearing were to be doubled, to 908 kg (2000 lb), the L_{10} life would be reduced to 11.7 million cycles.

There are some applications in which greater reliability is required and the typical L_{10} life is not adequate. To specify larger, more reliable bearings for those applications, the following multipliers would be used:

Desired Reliability Level	BDR Multiplier
L_{10}	1.0
L_5	1.6
L_4	1.9
L_3	2.3
L_2	3.0
L_1	4.8

HERTZIAN FATIGUE, ROLLING ELEMENT BEARING LUBRICATION, AND SURFACE FATIGUE

All the aforementioned formulas are based on the assumption that the bearing will fail from Hertzian fatigue. Heinrich Hertz developed the equations for expressing the fatigue stresses experienced by a plate exposed to a load from a rolling element about 150 years ago, and all rolling element bearing lives are based on an understanding of these stresses.

Figure 10.3 shows a typical ball or roller as it goes along its path. Looking at this diagram, one can see that, as the rolling element proceeds along the race, the surface elastically depresses in reaction to the compressive loads and this depression results in Hertzian stresses within the part. Fatigue failures cannot occur with purely compressive loads, but what happens within the bearing race is that the areas immediately before and after the depression are subjected to tension stresses. It is the tension in these areas, followed by the compression that results in the Hertzian fatigue cracking that all rolling element bearings are designed around.

A Hertzian fatigue failure begins with a fatigue crack that grows parallel to and about 0.1 mm (0.004 in.) below the surface, and then rapidly works its way to the surface. However, in reality, most of the industrial bearing failures we've seen have been from surface fatigue, cracking that starts on the rolling surface and is aggravated by contamination, dimensional errors, and poor lubrication. In Figure 10.4, the differences between how the cracking begins can be clearly seen.

THE REASONS BEARINGS FAIL

In the chapter on lubrication, we discussed how viscosity conversion works to support the bearing element in operation and this is a good time to reiterate:

- The typical rolling element bearing lubricant film thickness is on the order of 0.0008 mm (0.00003 in.).
- The pressures are frequently in the order of 1,400 MPa ($\approx$200,000 psi) but can go above 2,000 MPa in very heavily loaded bearings.
- 0.04% water in the lubricant (4 parts water in 10,000 parts oil) has been shown to reduce bearing life by a factor of 5.

According to bearing company data, most bearings are replaced because of wear but it is unusual that a failure analyses would be conducted on a worn rolling element bearing. Instead, we find badly damaged and occasionally melted pieces of metal and sometimes its hard to find all of the pieces. From our analyses, it appears that the major causes of bearing problems result from ignorance of the three statements in the preceding paragraph and we'll look at them in more detail in the following text.

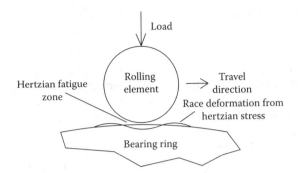

FIGURE 10.3 As the element rolls across the surface Hertzian fatigue stresses are created in the ring.

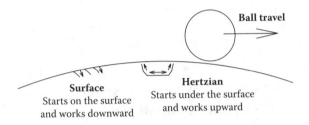

FIGURE 10.4 Bearing design is based on Hertzian fatigues, but they often fail from surface fatigue.

The typical rolling element bearing lubricant film thickness is in the order of 0.0008 mm (0.00003 in.). This lubricant film acts to spread the load over a greater area and reduce the stress concentrations in the element path. Dirt particles larger than the film thickness, water, and inadequate lubrication act to interrupt the film and increase the fatigue stresses.

The pressures are frequently in the order of 1,400 MPa (≈200,000 psi) but can go above 2,000 MPa in very heavily loaded bearings: The stresses caused by normal operation are tremendous. If we add to that the stress from an out-of-round bore, or a bore that doesn't properly support the shaft, and/or the vibration forces from a loose shaft fit, the life of the bearing is reduced. Table 10.1 is a listing of typical internal clearances for deep groove type ball bearings. Reviewing it, one can see that, for a normal bearing with a 50-mm (2-in.) bore, the internal clearance can range from 8 to 28 μm (0.00035 to 0.0011 in.). The machining specifications for the bore fit call for the maximum allowable out-of-roundness to be less than 14 μm (0.0005 in.).

TABLE 10.1
Typical Radical Internal Clearance for Ball Bearings

Bore Range		Typical Radial Internal Clearance					
Millimeters		Metric (min/max in μm)			Inch (min/max in 0.0001 in.)		
Over	Up to	Normal	C3	C4	Normal	C3	C4
30	40	6/20	15/33	28/46	2/8	6/13	11/18
40	50	6/23	18/36	30/51	2.5/9	7/14	12/20
50	65	8/28	23/43	38/61	3.5/11	9/17	15/24
65	80	10/30	25/51	46/71	4/12	10/20	18/28
80	100	12/36	30/58	53/84	4.5/14	12/23	21/33

From the preceding text, it is easy to understand how a 0.5 mm (0.020 in.) "soft foot" or the deflection of a weak base could distort the bearing housing and the outer race and greatly reduce the bearing life.

> The distortion caused by soft foot should not be underestimated. Several years ago we started what was intended to be a casual failure analysis on an inexpensive cast-iron two bolt pillow block. Analyzing the ball path, we were surprised to find there were two load zones, one centered at about 4 o'clock and one at about 8 o'clock. Our initial reaction was to think the person didn't properly shim the pillow block, but we eventually realized that the inexpensive fan base was deflecting under load, causing distortion of the housing. Since then more than 30% of the two bolt pillow blocks we've looked at have shown bore distortion problems.

As with many of the other failures we've mentioned, we frequently see more than one cause, and it is not unusual to see an application with an incorrectly installed bearing with contaminated lubricant in a loose fit — and the plant personnel are blaming the machine manufacturer for the short bearing life.

A DETAILED ROLLING ELEMENT BEARING FAILURE ANALYSIS PROCEDURE

The following procedure uses nine basic steps to be followed in looking for the causes of a failed bearing, and within each of the steps are numerous questions to be asked. It starts with an inspection of the housing and surroundings so as to better understand the bearing's operating conditions, then looks at the exterior of the bearing (to understand the heat transfer and supporting forces) and, then, the internals (to see the actual operating forces).

The nine steps we use in finding the physical causes of the bearing failure are:

1. Understand the background, i.e., the bearing's history.
2. Inspect the contact surfaces of the inner and outer rings, i.e., the inner- and outer-race fits.
3. Next, look at the sides of both ring exteriors.
4. Remove the seals and/or shields and visually examine the lubricant.
5. Look at the application, determine the loads, understand what the ball and roller paths should look like, and separate the rings in such a way that those paths can be inspected. Then place the rotating ring on a "lazy Susan" and slowly rotate the bearing while watching the ball/roller path.
6. Continuing with the physical inspection, place the fixed ring on a lazy Susan and slowly rotate it while watching the ball/roller path.
7. Test the rings and rolling elements for hardness.
8. Inspect the cage, paying careful attention to the surfaces that the elements come into contact with.
9. Inspect the rolling elements.

In the following text we've added a group of questions to the nine points, which, when they are answered, should lead to the physical roots of the failure. In listing these questions, we've tried to make comments about the various answers and why one or the other may be of importance. We've also included photographs illustrating many of the points. There will be times when some of the answers won't necessarily point to the bearing problems. They are used to help the investigator develop a better sense of the operation. However, deviations from what is normal and desirable are clues to failure causes and a starting point for further investigation.

Reading the following steps will be confusing the first time through because there is a lot of text and there are many questions. When that happens, go back and look at the list of the nine questions and try to patiently think about what you have seen. What we have tried to do is write down how we go about looking at a failure and how to interpret what is seen. Like the shaft failures mentioned earlier, the parts will tell you how they died. The difficult part is interpretation of the symptoms, but with patience and practice, it becomes much easier.

STEP 1

Understand the background, i.e., the bearing's history.

a. Look at the housing and the bearing's surroundings. Are they clean and uncontaminated? Some thoughts about the surroundings are:
   - Bearings are not perfect, and they generate heat. If debris is covering the housing, it will tend to increase the temperature of the parts inside. In turn, increased internal temperatures will lead to reduced lubricant viscosity, reduced film thickness, and more rapid additive deterioration. The bearings and gears in the conveyor drive in Photo 10.4 are being destroyed by their environment.
   - Water in lubricants is deadly to bearing life, and if the housing has been wet, there is the possibility that the lubricant has been contaminated.

b. Measure the bearing housing bore and the shaft dimensions in at least three positions on each side of the contact. Compare these readings with the specifications. After the basic lubrication is supplied, probably the most important single characteristic of a bearing installation is the dimensional fit. During operation, the elastic flexure of the bearing steel generates heat and most of that heat is usually conducted out of the bearing through the shaft and housing fits. If those fits are not sound, the internal bearing temperature increases and the clearances decrease. The heat generated is approximately equal at both rings and invariably the inner ring is smaller and has less mass, so it gets hotter and grows more. When the clearances decrease, the internal stresses increase. In addition, as mentioned immediately above, the increased temperature reduces the viscosity of the lubricant, and when viscosity conversion takes place, the resultant film is thinner and the fatigue stress in the races is increased.

PHOTO 10.4 The dust on this motor/reducer is destroying the bearings and the flimsy base is aggrevating the problems.

c. What is the range of temperatures and relative humidity readings in the area? Is the shaft hotter than the bearing?

 • If the shaft is hotter than the bearing, the inner ring is going to grow much more than the outer ring, and problems outlined in 1b could occur. If this is the case, what internal clearance is specified? (Motor bearings are usually C3 clearances because of the shaft runs hot. C4 bearings have even more internal clearance.)

 • The comment about relative humidity is directed at the possible effects of water contamination in the lubricant. The aforementioned data shows how water decreases bearing life.

d. When was the bearing installed? How fast does it rotate? Is the bearing application correct for the expected speeds and loads? How many hours has it run? Are the operating conditions the same as the machine was designed for? These questions go back to the design of the bearing and its application — so the analyst has a good idea of the design conditions and can compare them with the operating conditions. Things to keep in mind at this stage of the analysis include:

 • Ball bearing life is inversely proportional to the load to the third power, whereas for roller bearing life the factor is 3.3.

 • Bearing design is for a given number of revolutions and the primary effect of speed is that those design revolutions will be used up faster.

 • There are specified limiting speeds for bearings and they will depend on the lubricant. (However, if the application exceeds the maximum speed, most of the bearing manufacturers do have alternative surface finishes that allow higher speeds.)

e. What is the lubricant, what is the relubrication schedule, how is the relubrication accomplished, and what education have the oilers had?

 • Is the bearing being lubricated with grease or an oil?

 • When you look at how often the unit is being lubricated, is the real purpose of that relubrication to resupply the rolling elements with lubricant or is grease being used as an auxiliary seal?

 • How much grease is being pumped into the bearing, and how is it applied? There is data showing that a substantial percentage of bearing failures are hastened by overlubrication. Figure 10.5 is a relubrication guide that can be used to determine the relubrication frequency and lubricant quantity for most bearing applications above 100 r/min.

Photo 10.5 shows a shielded motor bearing that has been destroyed by a zealous but untrained oiler. Shielded bearings can be relubricated if a measured quantity of grease is carefully pumped in by hand but destroyed bearings with dimpled shields such as these can only be produced by an uneducated finger on the trigger of an automatic grease gun. Continuing on the thought of proper lubrication, although this photo shows a shielded bearing where the failure was caused by poor practices, damage can also be caused by overlubrication of open bearings. With overlubrication the steps in the typical failure involve:

1. The oiler pumps excessive grease into the bearing cavity.
2. Then the rolling elements have to continually push the lubricant out of the way and, as a result, generate more heat than normal.
3. The excessive heat not only causes reduced internal clearances and increased parasitic loads but also it reduces the lubricant film thickness.
4. The result of the excessive loads and the reduced film thickness is greatly reduced bearing life.

f. Review the maintenance records, including the vibration analysis and/or oil analysis history on the machine. In analyzing this data, is there a pattern showing a relationship between the first measurable deterioration and other plant activities?

g. Speak with the operating, maintenance, and engineering personnel involved with the machine. Ask for their ideas on what could have contributed to the problem.

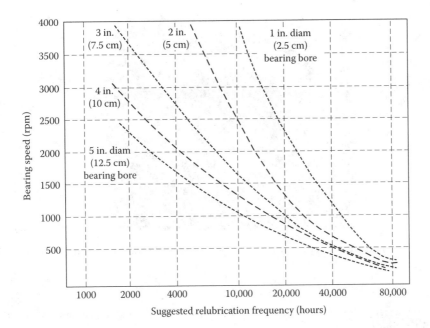

For the correct relubrication interval, multiply the relubrication hours value found above
by the correction factors shown below

Condition	Typical operating condition	Factor
Housing temperature F_t	Below 150° F	1.0
	150° to 175° F	0.5
	175° to 200° F	0.2
	over 200°	0.1
Contamination F_c	Light nonabrasive dust	1.0
	heavy nonabrasive dust	0.7
	Light abrasive dust	0.4
	heavy abrasive dust	0.2
Moisture Fm	Humidity almost always below 80%	1.0
	humidity between 80% and 90%	0.7
	occasional condensation	0.4
	occasional water on hottsing	0.1
Vibration F_v	Less than 0.2 ips peak velocity	1.0
	0.2 to 0.4 ips	0.6
	above 0.4 ips	0.3
Position F_p	Horizontal bore centerline	1.0
	45° bore centerline (to horizon)	0.5
	vertical centerline	0.3
Bearing type F_d	Ball radial and thrust bearings	1.0
	cylindrical and needle roller bearings	0.5
	tapered, spherical and others	0.1

Do not use
below 100 rpm!

Do not use
below 100 rpm!

Notes:

1. To calculate the best relubrication interval, start with the rpm and travel over to the shaft diam.
   Then drop vertically to the recommended relubrication frequency hours. Next, multiply these
   hours by each of the six charted correction factors.
2. The values shown in the table above represent the internal needs for lubricant. Often
   relubrication is used to act as a secondary seal and that is not addressed by this chart.
3. The relubrication intervals are based on the use of a good grade of non-EP bearing grease.
4. This chart can be used for ball thrust bearings but not for other types of thrust bearings.
5. The relubrication quantity can be calculated by
   Grease volume (fl. oz.) = 0.1 x Brg width (in.) x Brg OD (in.)

(c) Aug 2005 by SS&A

FIGURE 10.5 Chart for relubrication of rolling element bearings.

PHOTO 10.5 A motor bearing where an overzealous oiler with an electric grease gun collapsed the shield.

STEP 2

Inspect the contact surfaces of the inner and outer rings, i.e., the inner and outer ring fits, and answer the following questions.

a. Are there any black stains on the rings?
   Yes — There has been water present.
   No — Water has not contaminated the bearing.
   Comment — We know the critical effect water has on bearing life. The water contamination typically causes black stains on the rings or in the ball path as shown in Photo 10.6. The photo shows the actual ball path damage and the presence of stains such as these or the ones shown in Photo 10.7 are proof that water has been present.

> The first time I really recognized the importance of those black spots caused by water was early in my career when we were trying to understand why the most expensive bearings in the plant were failing at a greatly increased rate. One day, we were inspecting one of the failed bearings with a representative from the bearing company when he pointed out the black corrosion spots. They didn't look or feel like much, and we asked if we couldn't just take fine emery cloth and polish them out. The representative went on to say we could do that, but it wouldn't help the situation because the actual mechanism was:
>
> 1. Water gets into the bearing and the bearing corrodes.
> 2. The corrosion generates hydrogen, and some of the hydrogen atoms are absorbed by the steel.
> 3. The hydrogen atoms cause tiny cracks in the ring immediately below the corrosion spot.
>
> If we were to clean off the corrosion debris — the black oxide — it wouldn't do any good because the bearing steel already had tiny cracks in it. He went on to say that if we did run the bearing he felt we would be lucky if it lasted for 1/10th of the L_{10} life.

b. Other than water stains, is there any visible fretting or discoloration of the shaft or housing fit surfaces?

PHOTO 10.6 The two black lines are water spots (corrosion damage) on a ball path.

PHOTO 10.7 The black marks on the side of this inner ring are proof that water has been present.

Yes — The bearing has been moving in the housing or on the shaft.

No — The bearing fit was excellent.

Comment — Recognition of fretting is important, because it tells us that the bearing has been moving in the housing. Sometimes this movement is allowable, as with the floating bearing in a motor in which having two fixed bearings would lead to very short bearing life. But it is important to look at the installation and recognize when fretting occurs. Photo 10.8 shows a bearing where the design called for only one half to be supported and we can see:

1. The fit on the half that was supported was bad, as shown by the highly polished surface.
2. The other half is what a good fit should look like when the bearing is removed from use.

PHOTO 10.8 The design called for only ½ of this bearing to be supported — and the fit was very loose.

PHOTO 10.9 This dark discoloration is "fretting".

Photo 10.9 is an example of *fretting*. This bearing was loose in the housing and moved back and forth a tiny bit as the inner race rotated. The brownish-black material is the oxide that the movement created.

c. Is the original grinding pattern still visible over more than 50% of the surface?

Yes — Some fretting is acceptable. If most of the original bearing surface isn't attacked, the bearing race has been moving, but the fit is acceptable.

No — The movement is excessive, causing fretting and reducing the bearing's life.

Comment — As we mentioned in the preceding text, there are times when some fretting is inevitable. As long as it isn't excessive, the insulating effect is not substantial. If the vibration isn't substantial and if most of the original ground outer surface is visible, the fretting wasn't excessive.

d. Is the fretting over more than 50% of the fit surface, or is part or all of the surface polished from contact with the shaft or housing?

Yes – The looseness is so severe that the race is rotating in the housing or on the shaft. In a bearing operating at more than about 300 r/min heat transfer has been drastically reduced, and, as a result, the bearing loads are increased and life reduced. In addition, the movement results in vibration that increases the bearing load.

No — There is movement and it is reducing the bearing life by increasing the loads and the operating temperatures, but it is probably not critical.

Comment — As the fit worsens, the bearing tends to rotate more in the housing. As the ring spins, it becomes polished as shown in Photo 10.10. These are a pair of angular contact ball bearings that were installed in a reducer. The highly polished surfaces show that they were both relatively loose and turning in the housing. As a result, the

PHOTO 10.10 Both of these bearings were loose in the housing and the movement has polished the races.

PHOTO 10.11 The bearing has been polished, almost to a mirror finish.

internal clearances have been compromised, and the load isn't properly distributed on the ring.

Photo 10.11 shows the outer ring of a pump bearing that is so highly polished that reflections of the surroundings be seen in this mirror-like outer ring surface. Also, a careful inspection shows the reddish tinge on the central band is the result of fretting corrosion, and the upper edge of the race is turning blue from the heat generated within the bearing. This shows the race has been heated to about 250°C (500°F). From these clues, we know the lubricant film is nonexistent, and the internal loads are excessive.

e. Is there a black irregular surface on the housing or shaft fit, showing that the fit was extremely loose and the surface was heavily loaded?

Comment — If the fit is very poor and heavily loaded, the surface will frequently spall and show surface fatigue damage from the heavy contact forces. Photo 10.12 is of the bearing mounting surface on a roller shaft. The threads for the locknut are visible to the left, but these bearings were very loose on the shaft and the result was this black irregular damage.

Sometimes this surface fatigue damage will include regular patterns such as that shown in Photo 10.13. People will assign all sorts of causes to this, but it is only the repetition of the grinding pattern used in manufacturing superimposed on the loose fit. (This bearing came out of one of our vehicles, and a dimensional check showed it was highly unlikely that the original fits were what the bearing company specified.) A second interesting point that can be seen in this photo is that the side of the race is highly polished in a band about 6-mm (¼-in.) wide, from rotating against the shaft shoulder.

Before leaving the subject of shaft and housing fits, we have to recognize the possibility of ring cracking. A loose fit frequently leads to cracking of the outer ring because the piece is not well supported as the contact forces are applied and bending stresses are added to the other operating stresses. Photo 10.14 has a star-shaped crack toward the left side of the photo. The crack originated on the outer diameter of the outer ring of a

PHOTO 10.12 A horrendously loose fit.

PHOTO 10.13 Another very loose fit complicated by the grinding pattern.

PHOTO 10.14 A heavily fretted surface with a visible crack.

self-aligning pillow block used on a fan. Inspection of the ring found fretting on both the top and the bottom of the ring, showing the pillow block bore was not round and that there was considerable movement between the bearing and the housing. In addition, the black discoloration shows the bearing has been wet. This puzzled the plant personnel until they realized that the bearing was occasionally subjected to humid outside air while the housing was still around 15°C (≈60°F).

Similar fractures, because of multiple stresses, can happen on the inner ring, and Photo 10.15 shows the inner ring of a spherical roller bearing:

1. There is a crack across the ring.
2. The shaft mounting surface is heavily fretted, and there are polished circumferential bands that show it has been spinning on the shaft.
3. As with the bearing in Photo 10.13, the side of the race is highly polished, almost mirrorlike, from movement and contact against the shaft shoulder.

One last comment on the subject of shaft and housing fits is that occasionally, the initial appearance can be deceiving. In Photo 10.16, the fits are not great and, looking at the

PHOTO 10.15 The cracked inner ring of a through-hardened bearing.

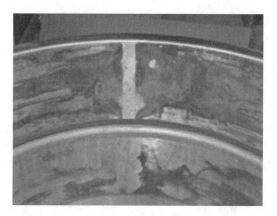

PHOTO 10.16 The fretting shows there was some movement, but the clear area over the keyway shows there was no rotation.

fretting on the inner ring surface of the two bearings, the suspicion is that they are very loose. However, the clean unmarked area that was over a lock ring slot shows that, although there was some motion, they were not rotating. These bearings had horrible outer race fits and badly contaminated lubricants. The inner race fit problem was a minimal contribution to their failure.

f. Are the seals and shields in good condition? If not, why not?

Comment — Seals and shields are designed to keep contamination out of the bearing, and even a small amount of water or dirt will reduce the bearing life because of increased surface fatigue stresses. Generally, lip seal life is only a fraction of the design life of a machine but, even after they degrade and lose contact, the seal lips tend to protect against water and larger particles. Inspect them; see if they were properly installed. Were they installed to prevent contamination from getting in, or were they supposed to prevent oil leakage out?

Photo 10.17 shows a view of a leaking seal lip at about a 10× magnification. The equipment manufacturer had been complaining about poor seal quality, but the circled area in the middle of the photo shows the seal lip was cut during installation at their plant.

PHOTO 10.17 The seal lip was cut during installation.

STEP 3

Next, look at the sides of both ring exteriors and answer the following questions.

    a. Does the original grinding pattern show clearly around the entire surface?
        Yes — The ring may have been moving against its restraint but the contact was not substantial.
        No — There has been movement and contact between the ring and the restraint.
    b. Are the sides of the rings polished? (See Photo 10.13 and Photo 10.14.)
        Yes — The ring had been moving against the restraint and the contact was substantial.
        No — There has been no movement or no contact.
    c. If there is polishing, is the pattern consistent with the mating piece pattern?
        Yes — That confirms looseness and relative movement.
        No — Differences in the patterns will show some unique motion is present and require detailed further investigation.
    d. Are there any black spots or stains on the sides of the rings?
        Yes — There has been free water present.
        No — Water has not contaminated the bearing.

STEP 4

Remove the seals and/or shields and visually examine the lubricant.

    a. Do the remains look discolored, contaminated, or oxidized?
        Yes — This shows the lubricant was not in good condition when the bearing failed.
        No — The lubricant did not play a substantial part in the failure.
        Comment — Analyzing the lubricant removed from a failed bearing is frequently challenging because the presence of deteriorated lubricant has to be weighed against the condition of the bearing at failure. One of the first questions that has to be asked is "Which came first, the deteriorated lubricant or the bearing failure?" It is extremely difficult to diagnose problems with bearings that have already had major component damage such as that shown in Photo 10.18. This bearing shows some of the classical symptoms of an inadequate lubrication failure, but how do we know that it didn't have other problems such as a poor fit, or brinelling, or improper metallurgy, etc., that have been hidden by the gross damage?

PHOTO 10.18

There will always be an argument between those persons who want to run a piece of
equipment until the last possible moment before a tremendous catastrophic failure
and those who want to shut the machine down and remove the part at the first indication of a serious defect. The first group tries to maximize their immediate output, whereas the second group, and I am one of them, tries to learn from their
mistakes. If we don't look at our errors and try to avoid them, what sort of progress
are we going to make on improving reliability? Trying to do a failure analysis on
a bearing such as that shown in Photo 10.18 is largely a waste of time.

---

b. Is there evidence of water?

   Comment — A crackle test will show only gross contamination but is relatively easy to
   perform. To perform the test, take a small sample of lubricant and drop it on to a hot
   plate that is at $160 \pm 5°C$ (320°F). If the sample crackles immediately after it hits the
   hot surface, it shows there is water in the sample. The crackling is caused by the boiling
   of water. This test is only accurate down to about 0.1% water, and we know that quantities below that can cause shortened bearing life, but it does give a good field guide.

c. Smell the lubricant. Has it been burned? Is there chemical contamination? Does it have
   a burned chemical smell or does it smell like the new grease that just came out of a
   grease gun? There are times when the lubricant will appear dry or powdered and this
   requires further examination. If the lubricant has been contaminated by dust or rust
   particles, these particles will frequently absorb any available oil. The appearance is that
   of an underlubricated bearing but many times the problem is contamination.

d. Take a small sample of the lubricant for a laboratory analysis.

   Comment — Frequently, a failed bearing will be blamed on inadequate lubrication or lack
   of lubrication, when, in reality, there was plenty of lubricant before the bearing got hot.
   After the failure there are times when all that is left is the residue from burned grease,
   and a careful inspection is needed to find the remains of any lubricant. On the other
   hand, there have been many times when a nervous oiler has added grease after the bearing has begun to fail. Be very careful during this phase of the investigation. If the bearing has failed with substantial damage and the grease looks like new, it is a good bet
   that it is and grease was added as, or after, the unit was shut down.

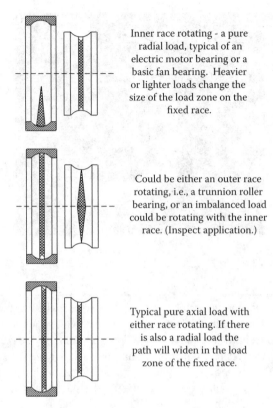

Inner race rotating - a pure radial load, typical of an electric motor bearing or a basic fan bearing. Heavier or lighter loads change the size of the load zone on the fixed race.

Could be either an outer race rotating, i.e., a trunnion roller bearing, or an imbalanced load could be rotating with the inner race. (Inspect application.)

Typical pure axial load with either race rotating. If there is also a radial load the path will widen in the load zone of the fixed race.

FIGURE 10.6 Typical normal load zones in a ball bearing.

STEP 5

Look at the application to determine the loads and understand what the ball and roller paths should look like, then carefully separate the rings so those paths can be inspected. Place the fixed ring on a lazy Susan and slowly rotate it while watching the ball/roller path.

    a. Is the ball/roller path where it was expected to be? If not, reanalyze the application to ensure your expectations were correct. Figure 10.6 shows some typical normal ball paths. Ball and roller paths and load zones should be relatively uniform and variations indicate the effects of loads. Some of the key points to understand include:
- If they wander from side to side and vary in width, other than that expected from the normal loads, there is misalignment, ovality, or dimensional errors in the application.
- Probably the first thing that should be done is to carefully measure the width of the ball or roller path, then measure the rolling element width or diameter.
  - There will be some variation with respect to load. (The heavier the ball is loaded the more it is deformed, and the wider the path.) In a lightly loaded application, the ball path should be about 0.2 times the ball diameter.
  - In a roller bearing the contact path should be the same width as the roller. Wider contact paths indicate dimensional variations.

> Using a lazy Susan for bearing inspection is one of the best diagnostic tools imaginable. For years we struggled with trying to inspect and interpret the ball and roller path variations. Now we just put the bearing on the lazy Susan, and spin it while varying the lighting and the parts tell us what happened.

Some abnormal ball paths

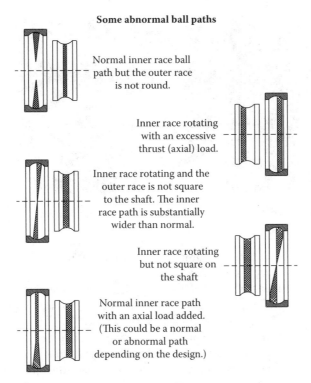

Normal inner race ball path but the outer race is not round.

Inner race rotating with an excessive thrust (axial) load.

Inner race rotating and the outer race is not square to the shaft. The inner race path is substantially wider than normal.

Inner race rotating but not square on the shaft

Normal inner race path with an axial load added. (This could be a normal or abnormal path depending on the design.)

FIGURE 10.7 Common abnormal ball paths in ball bearings.

Figure 10.7 shows some common abnormal ball paths.

b. Is there evidence of abnormal thrust loading, i.e., a path to one side of a ball bearing, polishing on the ribs of a tapered roller bearing, or unequal wear on the paths of a spherical roller bearing?

Yes — There is an axial (thrust) load.

No — There is no significant axial (thrust) loading.

Comment 1 — In a deep groove ball bearing, the ball path may be off to one side and these bearings are designed to carry some thrust load. But, looking at a cross section of the ring with the normal ball path at 6 o'clock, the path resulting from the thrust load should rarely be centered any higher than 7:30 o'clock.

Comment 2 — Analyzing the roller paths on spherical roller bearings is particularly important. The paths should appear essentially the same and uniform around the ring. These bearings are designed to take some thrust loading and there may be more wear on one of the raceways. But their primary load should be radial, and measurable wear on one path, with little or no contact on the other, is proof that the one path is carrying the entire load. Photo 10.19 shows the outer ring of a spherical roller bearing that was installed with no end float, i.e., the bearings were on a fan, and both bearings were fixed so they couldn't move when the shaft expanded and it is obvious from the photo that one race carried the entire load.

c. In a bearing with a pure radial load, how large is the load zone?

i. Less than 45° indicates a lightly loaded bearing.

ii. Over 120° indicates a very heavily loaded bearing.

iii. A load zone that does not gradually and uniformly grow larger and then smaller indicates geometry problems.

PHOTO 10.19 This fan bearing shows a heavy thrust load, the result of assembly errors.

d. In a bearing with a predominant thrust load — is the ball/roller path of consistent width and position around the entire race?
   Yes — Good. The load has been uniformly applied.
   No — If it varies in width, the question has to be asked, "Is it misalignment or is it the effect of expected loads from the machine and gravity?"
e. Look at the ball/roller path. It should be smooth and lightly polished, sometimes even hard to see compared to the very finely finished machined surface of some bearing races. (If it is hard to see, when you rotate it on the lazy Susan, shift the light to shadow the ball path from different angles.) However, a few of the common problem symptoms are:
   i. Black marks or spots — Indicate that water or corrosives have been present.
   ii. Highly polished — This mirrorlike appearance is the result of skidding and may be caused by inadequate lubricant film and/or water contamination. Photo 10.20 shows a ball path that looks like a mirror, and a careful inspection shows my digital camera. This was a bearing that was started without a lubricant. It was greased after about 3 h, but the damage was already done. The wide ball path is a symptom of other problems.

PHOTO 10.20 The mirror finish on this ball path is the result of inadequate lubrication.

PHOTO 10.21 This ball path is easy to see because contamination has damaged the surface.

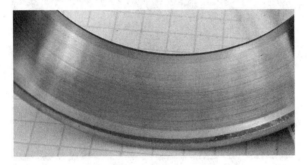

PHOTO 10.22 A heavily loaded and contaminated roller path.

iii. Matte surfaced — This damage results from either fine contaminants and debris in the lubricant or from grossly excessive loads. If the appearance is that of a light sandblast, the problem is usually contamination. If it appears heavily etched and blasted, the problem is excessive loads. Photo 10.21 is of a heavily loaded and badly contaminated ball path.

iv. Bruised — When hard foreign particles, such as debris from spalling, contaminate the surfaces and are rolled into the ball and race surfaces, they leave indentations. These indentations are called *bruises*. If the damage happens during the manufacturing process, before the bearings are finish ground, surface imperfections are called *dents*. The dings and damage in the upper roller path of Photo 10.19 are bruises.

v. Circumferential lines — Evidence of grit particles dragged along the rolling element path as shown in Photo 10.22.

vi. Spalled — A spalled surface is one where fatigue cracks have developed and pieces of the race have flaked loose. In the following text, we list and discuss several of the more common spalling appearances. But before going on to them a quick look at the spalls in Photo 10.23 is in order. This photo shows two spalls and a ball path that is well centered, heavily loaded, and both bruised and highly contaminated. We can tell from the shape of these spalls that they are not just the result of Hertzian fatigue, and we suspect other causes are present.

- *Single small deep spall* — Typically the result of Hertzian fatigue, arcing from high amperage electrical current passage, or a single external blow. Photo 10.24 shows a Hertzian fatigue failure of a roller bearing. The spall began below the surface on the far right and grew to the left as pieces flaked out of the surface.

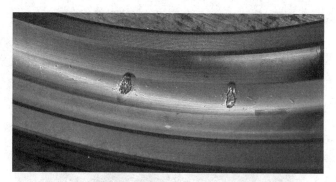

PHOTO 10.23 These two spalls are located on the ball centers.

PHOTO 10.24 A Hertzian fatigue spall in a roller bearing.

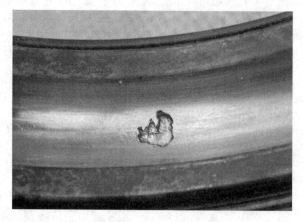

PHOTO 10.25 Another Hertzian fatigue spall.

Photo 10.25 also shows a Hertzian fatigue spall but this one is in the outer ring of a deep groove ball bearing. For both of these bearings, a major clue that they are from Hertzian (end-of-life) fatigue is the relatively deep nature of the spalls.

- Multiple spalls — If there are multiple spalls with same spacing as rolling elements, the cause is almost always either brinelling or false brinelling, but there will be an occasional failure caused by electric arcing when the current flow is high enough to simultaneously arc through at several elements. Three photos showing this uniform spacing are discussed in the following text:

PHOTO 10.26 A forklift truck spindle that has been brinelled by careless assembly procedures.

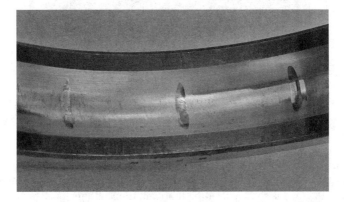

PHOTO 10.27 This false brinelling resulted from long-term vibration while the bearing was not rotating.

- The first, Photo 10.26, is a forklift truck spindle in which the mechanic has misaligned and driven the bearing onto the shaft. When a bearing is brinelled, the impact of the rolling elements being driven into the ring results in a series of dents that rapidly lead to multiple spalls.
- Photo 10.27, is a view of a bearing with false brinelling. Brinelling is a denting process that happens in a single blow. False brinelling happens over time and is actually a welding and corrosion process that removes material from both the ring and the elements. In this example, the outer ring has multiple depressions from false brinelling; if it had continued to run, it would have failed rapidly.
- Photo 10.28 is of a pump bearing that shows advanced damage and when a bearing has run for a while, it is almost impossible to tell the difference between brinelling and false brinelling. In this example, we dropped a pump while rigging it into position. It fell about six feet and landed on the shaft, brinelling the bearing. (We didn't have a spare and vibration analysis on the unit immediately indicated a problem. The pump ran for about nine months before it was removed from service.) The photo shows the regularly spaced damage that started with the brinell marks but has been multiplied by water contaminating the lubricant, resulting in many shallow spalls.

PHOTO 10.28 A brinelled bearing that ran for 9 months.

PHOTO 10.29 This "peeling" is frequently seen when the lubricant is inadequate.

- Continuous fine shallow spalled path — The result of an inadequate lubricant film, either the load is too great or the lubricant didn't do its job. This is also sometimes called *peeling*, and is an example of surface fatigue, in which the spall starts on the raceway surface and grows inward.
  - Photo 10.29 shows one path on a double-row spherical roller bearing, and the fine peeling is typical of a bearing with a serious lubrication problem. In this example, the lubricant was contaminated and couldn't do the job.
  - Photo 10.30 is actually the same bearing as seen in Photo 10.20, where the excessive end thrust overpowered the lubricant.
  - Photo 10.31 is a good comparison of the differences between the shallow surface fatigue on the left and deeper Hertzian fatigue on the right.
  - Irregular patches of spalling — A combination of forces. Photo 10.32 shows a combination of false brinelling (or brinelling) and lubricant breakdown. It could also have been one of them combined with shaft or bore eccentricity. In addition, notice that there was no load on the other ball path.
  f. Color — Carefully look at the color of the race and the ball path. Is the race the same color as when it was new? Using the reflection of a cool white fluorescent light, tilt the race and watch the reflection travel across the rolling surfaces. Is there any tint or change in surface color?
  - A pale gold tint indicates the rolling path has been over 400°F.
  - A blue color indicates the race has been well over 550°F.

PHOTO 10.30 More surface fatigue from an overloaded lubricant.

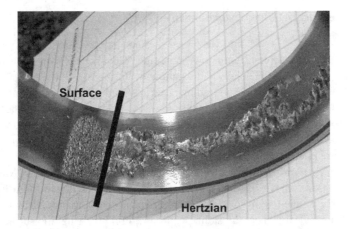

PHOTO 10.31 On the left is surface fatigue while the right shows the deeper Hertzian fatigue.

PHOTO 10.32 This ball path shows a combinatin of problems, probably brinelling combined with an inadequate lubricant.

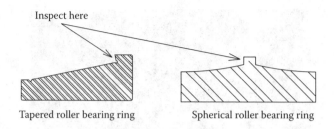

Inspect here

Tapered roller bearing ring   Spherical roller bearing ring

FIGURE 10.8 Critical places to inspect for evidence of thrust loading on roller bearings.

    g. Condition of thrust shoulders — With roller, spherical roller, and tapered roller bearings, inspect the condition of the thrust shoulders. (Figure 10.8 shows the critical areas to look at.) If they have seen contact, the surfaces should be lightly polished, not heavily burnished or mirror finished from sliding contact.

Appearance of Electrical Damage Mechanisms

Before we go on to Step 6, we should discuss the effect of electrical currents on the ball path appearance. A common occurrence with low-amperage, high-voltage current leakage across the element path is *fluting*. Photo 10.33 shows an example of fluting in a 266 kw (200-hp), 1800-r/min motor bearing. The individual lines are caused by electrical arcing and remelting of the bearing steel.

A close-up of some fluting damage can be seen in the magnified view of Photo 10.34. This is from a spherical roller bearing on a reducer, and one can see that the flutes are actually areas in which the surface has been removed. (One additional interesting point in this bearing are the corrosion pits in the upper half of the photo.)

There are times when mechanical fluting is confused with electrical fluting. Mechanical fluting results from periodic or constant pulsing loads, such as the load delivered to a bearing from the pounding of a poorly meshed spur gear. The bearing surface sees this persistent hammering, and if the hammering always happens at the same positions around the ring, the surface eventually can be plastically deformed. Photo 10.35 shows an example of mechanical fluting removed from a large reducer. The faint dark lines across the roller path are the flutes.

Without metallurgical analysis, differentiating between the two types of fluting is difficult. Electrical fluting is generally better defined and there are many more flutes, but a metallurgical analysis can give a definite answer. If electrical fluting is present, the damage will show as a white remelted area.

Photo 10.36 shows second type of electrical damage that is usually seen as a result of improper grounding during welding. The typical situation involves a welder not grounding close to the weld

PHOTO 10.33 "Fluting" from electrical discharge in a variable speed motor.

PHOTO 10.34 A magnified view of electrical fluting showing that material has been removed (approx. 5×).

PHOTO 10.35 Mechanical fluting of a roller bearing from a steel mill drive.

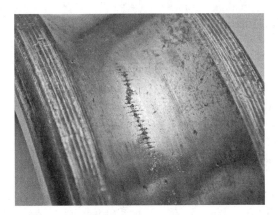

PHOTO 10.36 This pump bearing was rotating when the welding arc left this tell-tale track.

and a second ground path allowing flow through the bearing. The crew was welding on a pipe connected to a running pump and the resultant arced pits show each pulse of the AC welding current. Two notes on this bearing are as follows:

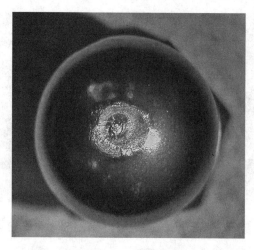

PHOTO 10.37 This is typical of the damage seen in a bearing ball after a high-current arc (such as welding).

1. This type of damage can also occur during other high-voltage, high-amperage electrical abnormalities.
2. This bearing also shows contamination and corrosion damage.

If that welding had been done while the machine wasn't running, it is likely that the elements would have been welded to the rings, and we have seen several examples of that.

Photo 10.37 is of a bearing ball that has been subjected to high currents. As with the fluting damage, positive proof of this can be found through a metallurgical examination.

> One expensive example of welding current being transmitted through the rolling elements occurred on a large crane in a wood yard. The plant scheduled some repair welding on the crane boom, and the welders hooked their ground clamp to the base of the crane while pulling the electrode 100 ft in the air to the boom. They completed their welding and the operator mentioned that it was tough to get the crane to swing the first time after they were done. Several weeks later he mentioned that the crane swing was rough and "grumbling." An inspection of the 2-kg (4.4-lb) bearing balls found that about half of them had suffered arcing damage from the welding.

STEP 6

Continuing with the physical inspection, place the rotating ring on a "lazy Susan" and slowly rotate it while watching the ball/roller path.

Ask the same questions that were asked in Step 5.

STEP 7

Hardness test the races and rolling elements.

Comment — All of the bearing manufacturers we know of state that the rolling element and race hardness should be between HRC 58 and HRC 62. Many of them also say the rolling elements should be a point or two harder than the races, but this frequently isn't true. (If the pieces are softer than HRC 58, the reduction in hardness is a time/temperature reaction that can frequently be used to understand more about the failure progression. Bearing steels start to lose their hardness when they exceed 200°C (400°F), and the longer and hotter the temperature, the more the steel softens.)

Step 8

Inspect the cage, paying careful attention to the surfaces the elements contact.

   a. Has there been contact on both sides of the element pockets?
      Yes — This is a situation that requires a bit of thought.
      i.   In a pure radial load, the rolling element goes into the load zone and drives the cage. As soon as the load zone ends, the cage drives the element. Therefore, there may be some light wear on both sides of the pockets but there shouldn't be any plastic deformation. You should not be able to feel a ridge of material along the top of the element pocket.
      ii. The same holds true in a reversing application.
      iii. However, if there is substantial wear, plastic deformation of the edge of the cage, or a cage fracture, it indicates there are either substantial reversing loads or poor lubrication. (These reversing loads generally occur only with torsional vibration and are a common problem source.) In Photo 10.38 this angular contact ball bearing cage shows tremendous plastic deformation resulting from two significant problems: the bearing was so heavily loaded that the lubricant film broke down and there was a serious torsional vibration in the drive.
      iv. In a bearing with a combined load, i.e., a combination of radial and axial loads, there should be wear on only one side of the pockets. If there is wear on both sides, it indicates there are definitely torsional vibration problems.
      No — This should be the case with most bearings.
   b. In a roller bearing, is the cage wear uniform across the cage bars and around the entire cage?
      Yes — The geometry of the bearing is acceptable and the loads are consistent with good design.
      No — The bearing rollers are not rolling correctly. Either the lubricant film is inadequate or the geometry is incorrect. 10.39 shows a spherical roller bearing cage. If the bearing had experienced pure radial loads, the cage might be slightly worn on the cage bars. But this cage shows heavy wear on both sides of the pocket as well as wear on the near side of the pocket. This combination of symptoms shows there are serious lubrication problems and possible reversing loads during each rotation.

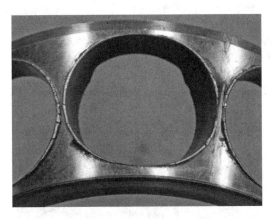

PHOTO 10.38 A bronze cage showing plastic deformation resulting from torsional vibration and marginal lubrication.

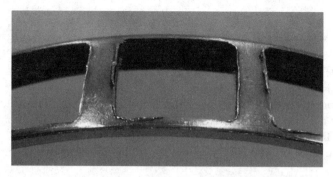

PHOTO 10.39 The cage wear has the same causes as that shown in Photo 10.38.

    c. In a ball bearing, is the cage wear uniform around the entire cage?
       Yes — The geometry of the bearing is acceptable and the loads are consistent with good design.
       No — The balls are not rolling uniformly during the rotation and the bearing's operating geometry is incorrect.

STEP 9

Inspect the rolling elements.

For Ball Bearings:

    a. Is the ball surface uniformly polished with a fine ground surface?
       Yes — Good. This is indicative of good operating conditions and satisfactory lubrication.
       No — The common conditions that contribute to rolling element damage are:
         • Black spots and/or blackened grooves — The black stains are corrosion products and indicate water has been present when the bearing wasn't rotating. The black smears on the balls in Photo 10.40 are proof that water was present.
         • Highly polished mirrorlike surface — The result of skidding and a poor lubricant film. Frequently evidence of water during operation.
         • Small circular scratch marks — The result of hard, relatively large contamination particles that were embedded in the cage.
         • Matte, lightly sandblasted appearance — Three likely causes are
            • Fine contamination in the lubricant.
            • High-frequency, low-amperage electrical discharges.
            • Extremely heavy loads breaking down the lubricant film.

PHOTO 10.40 The black spots on these balls show water has been present.

PHOTO 10.41 The lines on these balls show they have not been rolling freely and there are dimensional problems.

>   Comment — A lubricant wear particle analysis should be able to tell if the lubricant was contaminated. However, both of the first two causes will result in fine-particle contamination. A simple metallurgical examination in which the race is cut and the cross section polished and etched should reveal if there is any rehardening from electric arcing.
> b. Are there obvious bands or stripes around the balls?
>   No — Good. The elements should roll freely, and there should be no pattern of scratches or circular bands.
>   Yes — This could be the result of several conditions as follows:
>   • Fine circular scratches — These usually result from hard debris particles caught between the ball and the cage.
>   • Discolored circumferential bands — typically, the result of inadequate clearance between the inner and outer races. The bands form when the balls can't freely rotate as shown in Photo 10.41.

For Roller Bearings

Comment — The contact surface of a roller is much larger than that of a ball and should be a straight line. As a result, roller bearings place more severe demands on both the lubricant film and the shaft and housing geometry. For example, a slightly tapered fit that would result in an imperceptible minor shift in the ball path of a ball bearing will result in significantly increased stresses on a roller bearing.

> a. Is the roller surface uniformly polished with a fine ground surface?
>   Yes – Good. This is indicative of good operating conditions, good alignment, and satisfactory lubrication.
>   No — The common conditions that contribute to rolling element damage are:
>   • Black spots or patches — The black stains are corrosion products and indicate water has been present when the bearing wasn't rotating.
>   • Highly polished mirror-like surface — The result of skidding and a poor lubricant film. Frequently evidence of water during operation.
>   • Circular scratch marks around the rollers — The result of hard relatively large, contamination particles wedged into the cages.
>   • Matte, lightly sandblasted appearance — See earlier discussion on ball bearings.
> b. Do the ends of the rollers have circular worn bands, stripes, or small semicircular marks?
>   No — Good. The ends should be as originally ground or very slightly polished. They should not have a mirror finish or have a pattern of scratches or bands.
>   Yes — Highly polished roller ends almost always result from either:
>   • Improper (low) lubricant viscosity.
>   • Moderately elevated thrust loads that have destroyed the lubricant film and resulted in metal-to-metal contact.

PHOTO 10.42 The grooves and rings on these rollers show the thrust load has overcome the lubricant film.

PHOTO 10.43 The "semi-colons" on the end of this roller testify that very high thrust loads have occurred.

> Comment — Highly polished roller ends with circular score marks or semicolons on them as shown in Photo 10.42 and Photo 10.43, show that either substantially elevated thrust loads or very poor lubrication has occurred. Rollers are not made with pretty swirl patterns on them.

ROLLER AND TAPERED ROLLER BEARING MOUNTING SURFACES

The accuracy requirements of roller bearing shaft and housing fits are much more demanding than those of ball bearings. If a ball bearing ring fit varies by 0.02 mm (0.0008 in.) across the width of the bearing, the rolling path is shifted a miniscule amount toward one side and the raceway radius may vary a little. But in a roller bearing, the design calls for the load to be distributed across the entire ring, and misalignment can have a serious effect on the fatigue stresses. The magnitude of the effect is a function of the size and load and larger units are less affected, but any visible difference in the roller paths is cause for concern.

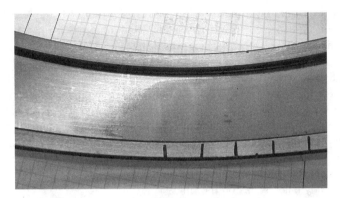

PHOTO 10.44 This race shows not only bore distortion but also mechanical fluting.

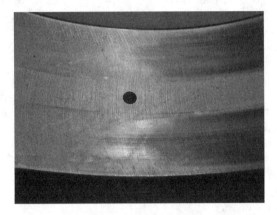

PHOTO 10.45 A well-distributed load showing signs of lubrication problems.

Photo 10.44 is of a roller bearing with the ring bore causing distortion and a heavier load on the side toward the reader. In this example, there is also mechanical fluting with the individual flutes indicated by the black marks we've made on the edge of the ring.

Photo 10.45 shows the roller paths from a spherical roller bearing that shows a pure radial load that was well divided between the paths and there was no measurable deformation of the outer ring. Unfortunately, the discoloration tells us that the raceway has been very hot and the highly polished surfaces show the lubricant film was nonexistent.

A DETAILED ROLLING ELEMENT BEARING FAILURE ANALYSIS PROCEDURE

As a guide, the nine steps are presented in the following text without the comments and illustrations:

1. Understand the background, i.e., the bearing's history.
   a. Look at the housing and the bearing's surroundings. Are they clean and uncontaminated?
   b. Measure the bearing housing bore and the shaft dimensions in at least three positions on each side of the contact. Compare these readings with the specifications.
   c. What is the range of temperatures and relative humidity readings in the area? Is the shaft hotter than the bearing?

    d. When was the bearing installed? How fast does it rotate? Is the bearing application correct for the expected speeds and loads? How long has it run? Are the operating conditions the same as the machine was designed for?

    e. What is the lubricant, what is the relubrication schedule, how is the relubrication accomplished, and what education have the oilers had?

    f. Review the maintenance records, including the vibration analysis and/or oil analysis history on the machine.

    g. Speak with the operating, maintenance, and engineering personnel involved with the machine. Ask for their ideas on what could have contributed to the problem.

2. Inspect the contact surfaces of the inner and outer rings, i.e., the inner and outer ring fits, and answer the following questions.

    a. Are there any black stains on the rings?

    b. Other than water stains, is there any visible fretting or discoloration of the shaft or housing fit surfaces?

    c. Is the original grinding pattern still visible over more than 50% of the surface?

    d. If the fretting is over more than 50% of the fit surface, is part or the entire surface polished from contact with the shaft or housing?

    e. Is there a black irregular surface on the housing or shaft fit, showing that the fit was extremely loose and the surface was heavily loaded?

    f. Are the seals and/or shields in good condition? If not, why not?

3. Next, look at the sides of both ring exteriors and answer the following questions.

    a. Does the original grinding pattern show clearly around the entire surface?

    b. Are the sides of the rings polished?

    c. If there is polishing, is the pattern consistent with the mating piece pattern?

    d. Are there any black spots or stains on the sides of the rings?

4. Remove the seals and/or shields and visually examine the lubricant.

    a. Do the remains look discolored, contaminated, or oxidized?

    b. Is there evidence of water?

    c. Smell the lubricant. Has it been burned? Is there chemical contamination?

    d. Take a small sample of the lubricant for a laboratory analysis.

5. Look at the application to determine the loads and understand what the ball and roller paths should look like, then carefully separate the rings so those paths can be inspected. Place the fixed ring on a lazy Susan and slowly rotate it while watching the ball/roller path.

    a. Is the ball/roller path where it was expected to be?

    b. Is there evidence of abnormal thrust loading, i.e., a path to one side of a ball bearing, polishing on the ribs of a tapered roller bearing, or unequal wear on the paths of a spherical roller bearing?

    c. In a bearing with a pure radial load, how large is the load zone?

    d. In a bearing with a predominant thrust load — is the ball/roller path of consistent width and location around the entire race?

    e. Look at the ball/roller path. It should be smooth and lightly polished, sometimes even hard to see compared to the very finely finished machined surface of some bearing races. If it is hard to see, when you rotate it on the lazy Susan, shift the light to shadow the ball path from different angles. However a few of the common problem symptoms are:

        i. Black marks or spots

        ii. Highly polished

        iii. Matte surfaced

        iv. Bruised

        v. Circumferential lines

      vi. Spalled
- Single small deep spall
- Multiple spalls
- Continuous fine shallow spalled path
- Irregular "patches" of spalling

    f. Carefully look at the color of the race and the ball path. Is the race the same color as when it was new? Using the reflection of a cool white fluorescent light, tilt the race and watch the reflection travel across the rolling surfaces. Is there any tint or change in surface color?

    g. With roller, spherical roller, and tapered roller bearings, inspect the condition of the thrust shoulders.

6. Continuing with the physical inspection, place the rotating ring on a lazy Susan and slowly rotate it while watching the ball/roller path. Ask the same questions that were asked in Step 5.

7. Test the races and rolling elements for hardness.

8. Inspect the cage, paying careful attention to the surfaces the elements contact.

    a. Has there been contact on both sides of the element pockets?

    b. In a roller bearing, is the cage wear uniform across the cage bars and around the entire cage?

    c. In a ball bearing, is the cage wear uniform around the entire cage?

9. Inspect the rolling elements.

For ball bearings:

    a. Are the ball surfaces uniformly polished with a finely ground appearance?

    b. Are there obvious bands or stripes around the balls?

For roller bearings:

    a. Are the roller surfaces uniformly polished with a finely ground appearance?

    b. Do the ends of the rollers have circular worn bands, stripes, or small semicircular marks?

SUMMARY

As with everything else that fails, there are almost always multiple causes. Bearings are extremely complex devices with multiple components machined to extraordinary tolerances, and one unfortunate result is that it is easy to accidentally overlook a major contributor to the failure. Sit down with another person and look at the nine points above. Carefully answering the questions will lead you to the solution.

11 Gears

The design and operation of gears is more complicated than almost any of the other common mechanical devices. They incorporate cantilever fatigue loading of the teeth similar to a beam in bending, Hertzian fatigue loading of the contact surfaces similar to a rolling element bearing, plus the lubrication demands of a pair of surfaces that involve both sliding and rolling. They are incredibly complex, and their failure analysis isn't for the faint of heart or for those whose primary athletic ability lies in jumping to conclusions. From the basic principles covered in the earlier chapters and careful inspection of the failed components, determining the source of most gear failures isn't terribly difficult. However, to help in their understanding, the first part of this chapter explains some of the more important details of gear design and function.

GEAR TERMINOLOGY

Gear operation involves both rolling and sliding contact. The first gears were designed so their contact involved only rolling, but that proved impossible to sustain in a real-world environment. As a result, almost all the gears we see have involute profiles, and as they contact they essentially slide into mesh, then involve more rolling as the teeth approach the pitch circle where the contact motion is pure rolling, and then as they go out of mesh, the high proportion of sliding movement returns.

(An *involute profile* is the shape traced by a point on a circle as the circle rolls along a surface. This gives us the typical curved shape of a gear tooth. There are other tooth profiles used, but none of them are in common industrial usage.)

There are dictionaries of gear terms and much of the terminology is truly specialized. To anyone but gear specialists, designers, and manufacturers, most of the terms generally aren't needed, and the ones shown in Figure 11.1 are by far the most common.

The gear terms used in this book are:

- *Active Profile* — The contact surface of the tooth. It includes the addendum plus most of the dedendum.
- *Addendum* — The top half of the tooth.
- *Backlash* — The effective clearance between the mating teeth.
- *Circular Pitch* — The distance between identical points on adjacent teeth, as measured along the pitch diameter.
- *Dedendum* — The lower half of the tooth. It is essentially the same as the addendum except the root clearance is added to it.
- *Gear* — The gear, or the *bull gear,* is the driven unit. It is usually, but not always, the larger of the two gears.
- *Pitch Diameter* (or *Pitch Circle*) — The ideal line the meshing gears intersect along. It is the ideal diameter of the gear.
- *Pinion* — The driving gear that is usually, but not always, the smaller gear.
- *Root* — The "almost flat" section between two adjacent teeth.
- *Root Clearance* — The clearance between the top land of one tooth and the root between the mating teeth.
- *Tip or Top Land* — The flat top of the gear tooth.

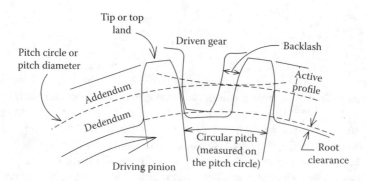

FIGURE 11.1 This diagram shows the gear terms used in this chapter.

One characteristic not shown in the figure is the *pitch angle*. This is the angle as measured from the tangent to the gear, on which the force is ideally transmitted. In the early 1900s, 14½° gears were by far the most common. As time went on, the teeth became shorter and fatter (and stronger in a bending mode) and 20° gears became the pressure angle of choice. Later, and in specialized cases, even higher pitch angles were sometimes used because the teeth were stronger and less likely to fail catastrophically. The downside of these higher pitch angles is that the separating forces between the gears become higher and the gear train becomes slightly less efficient.

TYPES OF GEARS

Looking back in history, the first "gears" were wooden devices in which the spokes of two hubs meshed with each other to transmit power and synchronize the equipment. As the industrial revolution began the idea of actual teeth working together was developed and these gears were straight-tooth *spur gears* as shown in Figure 11.2.

One of the difficulties with spur gears is that the power pulses are discrete, i.e., as each pair of teeth mesh a power pulse is delivered, and they are relatively noisy and rough. An early solution to this was the *helical gear*. With helical gears, there is generally more than one tooth in mesh, and the resultant power transmission is much quieter and smoother than with spur gears. The downsides of the helical gear are that there is a very slight decrease in efficiency and there is also a resultant thrust load. So the next step in gear development was to compensate for the thrust load,

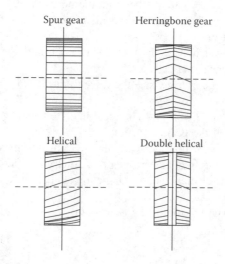

FIGURE 11.2 Four common gear designs.

PHOTO 11.1 A "cluster gear" from an automotive transmission.

and this resulted in the development of *herringbone* and then *double-helical* gears, both of which are thrust balanced.

(A practical example of the difference between helical and spur gears can be seen in the typical automobile. Driving the car in reverse produces a whining noise that doesn't happen in first gear. The gear ratios are almost identical, with the helical-first-gear operation is smooth and quiet, while the spur gear used for reverse produces a whining noise and vibration from the individual power impulses.)

Photo 11.1 shows a "cluster gear" assembly from a manual automotive transmission. There are the four helical gears used for forward speeds and the spur gear used for reverse. (A second interesting feature of this assembly is the difference in the size of the tapered roller bearings, partially a result of the point where the load is delivered but largely the result of the thrust force produced by the helical gears.)

Spur, herringbone, and double-helical gears have to be used with parallel shafts. Also, most helical gears are used with parallel shafts, but there are many other gears designed for use on nonparallel shafts. Two of these are shown in the photos and others are described in the text.

Photo 11.2 and Photo 11.3 both show gears typically used in right angle drives. The bevel gear is essentially a right-angle spur gear and is similarly rough and noisy. The spiral bevel gear, with multiple teeth in mesh, is similar to a helical gear in its smoothness.

Other common gear types include *hypoid gears*, which are variations on spiral bevel gears set up so their centerlines don't intersect, and *worm gears*, which involve pure sliding motion, as opposed to the rolling and sliding motion of the other gear designs.

PHOTO 11.2 A bevel gear — essentially a spur gear for right-angle drives.

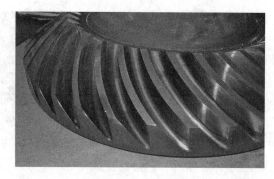

PHOTO 11.3 A spiral bevel gear — resulting in smoother power transmission than the bevel gear.

TOOTH ACTION

Before analyzing the stresses that occur within the gears, a review of the relative tooth movement and contact will make the wear and fatigue action much easier to understand. Three phases of this relatively complex contact are shown in Figure 11.3, and, as can be seen by the notes on the right side of the figure, the contact combines both rolling and sliding motions.

As the teeth first engage, the tip of the driven tooth slides down the driving tooth but at the same time the contact point rolls upward. The combined motion yields a low relative speed at the contact point, and frequently there isn't enough velocity to create an adequate lubricant film. As the tooth contact progresses, the contact point reaches the centerline, the tooth action is pure rolling in the upward direction, and with increased speed, a thicker lubricant film is developed. As the contact continues further along, the rolling and sliding motions are in the same direction and a substantial lubricant film is developed.

In addition, as the contact point slides downward along the driving tooth surface, those contact areas that are in tension are subjected to fatigue damage. However, at the same time, the contact areas of the mating tooth are in compression and fatigue doesn't occur on them. Conversely, later in the cycle, in the latter phases of contact, the roles are reversed and the driven gear dedendum can suffer fatigue damage. This combination of motions and fatigue stresses, i.e., the negative and positive sliding stresses, explains why damage such as pitting and surface wear is most often seen in the lower halves of the driving and driven teeth.

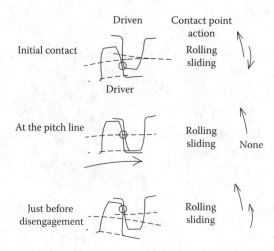

FIGURE 11.3 Three stages of gear tooth meshing and the relative motions involved.

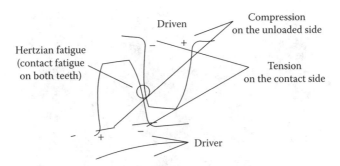

FIGURE 11.4 This diagram shows the typical fatigue stresses in gear teeth.

Not only are the teeth subjected to the Hertzian fatigue loads, they also see a cantilever bending load, and the combined loads are shown in Figure 11.4.

> In an interesting verification of the idea that fatigue cracks cannot occur in compression, we once tracked a pair of large cracks on the back side of a very heavily loaded cast steel through-hardened gear. The cracks were the result of manufacturing errors and did not have measurable growth over the years that we watched them, but the contact side of the teeth had substantial wear.

LOAD AND STRESS FLUCTUATIONS

Theoretically the power transmitted by the driving teeth should be a constant at some level, such as 100 hp, 500 kW, etc., and the teeth should all be loaded identically. But what actually happens during operation is that the forces fluctuate during each revolution. Some of the sources of these changes are:

- Coupling misalignment
- Gear misalignment
- Input power variations
- Gear and pinion eccentricities
- Machining errors
- Torsional vibration and resonances

The result of this is that a graph of the load on the teeth, instead of being a flat straight line, actually varies as shown in Figure 11.5. In this example the peak tooth load is almost twice the average load, and these variations are important because it is the peak loads that cause the bending fractures and Hertzian fatigue damage.

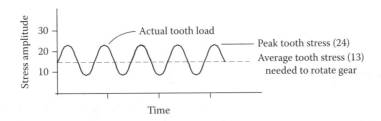

FIGURE 11.5 Comparing the actual tooth load vs. the ideal load in a typical gear drive.

PHOTO 11.4 The gear drives in only one direction. The wear on the "back" or inactive side of the teeth shows the peak drive load is more than twice the average load.

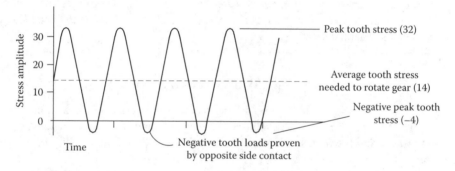

FIGURE 11.6 In a drive with torsional vibration the stresses are more than double the design loads.

Of special concern are those applications in which damage (wear or polishing) is seen on the "noncontact" side of gear teeth. The damage could be the result of misalignment (which will be dealt with later), but it could also indicate that there have actually been reversing loads and tremendous stresses on the teeth.

Photo 11.4 shows a 250 mm (10-in.) diameter cam drive gear from a 5300 kw (4000-hp) gas engine with the teeth worn on both sides. The gear rotates clockwise and is supposed to drive in only one direction, but the wear is positive proof of the reversing loads that occur within the gear train. Looking at both the gear and the values in Figure 11.6, it can be seen that the peak stresses resulting from reversing loads must be more than twice the average driving load and that serious damage can be occurring throughout the gear train. (In motor-driven gear sets we have found high-speed-recording ammeters and power meters to be invaluable diagnostic tools.)

GEAR MATERIALS

The first spur gears were made from wood, and there are still some low-speed and lightly loaded gears that have maple teeth bolted into cast-iron hubs. Today, most of the larger gears are metal, but polymers have made substantial inroads, and an interesting example of a blend of old and new technologies is the use of nylon teeth and steel hubs to quietly replace the cast-iron gears driving the dryers on older paper machines. Of the all-metal gears, most are steel; although there are cast-iron and bronze units in operation. Cast-iron gears are used almost exclusively on older machinery in which there are relatively light tooth loads, but they are also used on some specialized equipment such as large rotary vessels and filters. The largest industrial use of bronze gears appears to be on worm-gear applications, where hardened steel worms are commonly mated against bronze worm wheels.

Over the years, there have been changes in the metallurgy of steel gears and, as the metallurgy changes, the deterioration and failure appearances change. Therefore, in the failure analysis of steel gears, it is helpful to understand both the hardness and the metallurgy of the teeth.

The gears originally used in most industrial applications were through hardened and were designed to wear with age, but by the 1930s surface-hardened teeth began to appear. The terms can be confusing and need to be defined:

- *Through hardened* means that the hardness of the gear tooth is consistent throughout the tooth, even though the steel may not have been hardened at all. Most through-hardened teeth are relatively soft and almost all are less than HRC 40.
- *Case-* or *surface-hardened* gears have a very hard and wear-resistant outer surface with a much softer core, as shown in Figure 11.7. The surface hardness is almost always above HRC 50 and they are designed to show essentially no wear at all, even after years of operation.

(Therefore, in analyzing a wearing or a failed gear, the place to start is to understand the gear hardness, but this is sometimes difficult in a field application. In those cases we frequently check the tooth surface with a file. If the file skids, the steel is hardened and any surface damage more serious than mild micropitting is cause for concern.)

In the early industrial cast-iron-gear applications, the gears were frequently run without grease or oil and the graphite in the iron served as a lubricant. As time went on, it was realized that harder materials wear more slowly. However, the trend toward harder materials was slowed by the fact that not only does impact strength decrease as the hardness increases, but also harder gears require closer tolerances in manufacturing and operation to prevent breakage. However, a big advantage of surface-hardened gears is that they are smaller and lighter (and therefore less expensive) and can convey the same power in a much smaller package.

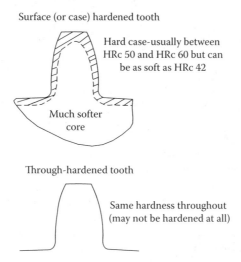

FIGURE 11.7 Comparing surface- and through-hardened gear teeth.

Most of the surface-hardened gears in industry are used in enclosed reducers. In the U.S. almost all of the reducers made before 1960 used through-hardened gears (that really were not very hard), but some of the European gear manufacturers began to build reducers with case-hardened gears shortly after World War II. By 1970 it was common to see American manufacturers making reducers with surface-hardened gears for use up to about 133 kw (100 rated hp). By 2000 most 1300 kw (1000-hp) reducers have case-hardened gears because the original cost is less and, if the unit is operated correctly, the gears should essentially last forever.

In looking at the capabilities of the different gear metallurgies, note that gear ratings are very dependent on geometric conformity and surface finish. Going to higher-quality machining, with greater precision and smoother surface finishes, allows more power to be transmitted within the same physical configuration. With this in mind, the performance differences between the two basic metallurgical structures are:

Surface-Hardened Gears
- Provide more power in the same size package
- Are used on almost all mobile equipment
- Demand greater precision in manufacturing
- Much less tolerant of surface damage and impact loads

Through-Hardened Gears
- Used on large open gear sets and reducers where precision is difficult
- Are much more tolerant of shock and impact loading
- Can tolerate substantial wear before failure

In addition, because they maintain their original geometry, case-hardened gears usually run quieter and slightly more efficiently than their worn through-hardened counterparts.

TOOTH CONTACT PATTERNS

In the preceding discussion the need for good alignment was mentioned several times. Looking further at the stresses between mating teeth, Figure 11.8 shows the ideal stress on a gear tooth and what actually happens as the contact pattern becomes poorer.

This shows the importance of good tooth contact, but in understanding it several points that should be mentioned are:

1. In the first panel of Figure 11.8 the mesh contact pattern is the actual contact pattern between the two teeth. We will assume it is as good as it looks and plot the stress as being uniform across the tooth.
2. In the second panel there is contact all the way across the tooth but it becomes weaker and just ends at the end of the tooth. In this example, in order for the driving force to be the same as the first panel, the maximum contact stress on the left side has to be twice the ideal (average) stress.
3. The third panel has the contact ending at midtooth. From this, the peak stress has to be at least four times that of an ideal tooth.

In analyzing gear tooth contact patterns, there are two further cautions that we should make. The first is, because of elastic deformation, the contact stress is frequently much more severe than it appears to the eye. For instance, if the teeth in the center panel of Figure 11.8 were inspected during a no-load operation, they might look like those of the third panel. However, because the

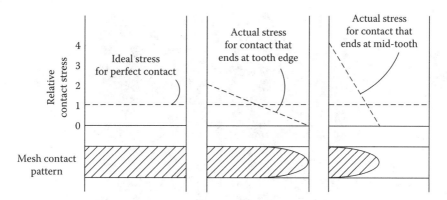

FIGURE 11.8 Comparing tooth contact patterns with relative contact stress.

teeth deflect under load, the apparent pattern is that seen in the second panel. The second caution is that housings of heavily loaded reducers frequently deflect because of the forces on the shafts. In these applications checking tooth contact patterns during a shutdown inspection can be misleading because the deflection force isn't present. Reducer inspections have to include well-lighted and very careful tooth inspections to be certain of the actual running patterns.

Photo 11.5 shows corrective pitting on the intermediate pinion of a large reducer. We had checked the gear tooth alignment while the unit was down and it was as perfect as humanly possible, yet in operation there was tremendous pitting on one side of the teeth. After a more detailed investigation, we found the 750 mm (30-in.). I-beam structure supporting the reducer was shifting by more than 4 mm (0.157 in.) when it was loaded. This was our first recognition of "dynamic soft foot."

With through-hardened gears, misalignment such as that in Photo 11.5 frequently appears as pitting after only a short time in run. With a surface-hardened gear, such as the haul truck drive pinion shown in Photo 11.6, there should be no apparent damage, and the contact pattern is more difficult to see. In this example the contact pattern that ends well before the edge of the pinion tooth shows that the peak contact and bending stresses are well above two times the ideal design stress.

The emphasis on these contact patterns is needed because they are tell-tale signs of the actual forces on the teeth during operation. Looking at the bending stress that causes tooth breakage, we

PHOTO 11.5 Corrective pitting of a through-hardened gear.

PHOTO 11.6 The contact pattern testifies to the misalignment of this case-hardened pinion.

see that it is a direct function of the load, i.e., if the transmitted power increases by 30%, the bending moment increases by 30%. However, the Hertzian fatigue loading is much more complex. For roller bearings in which the fatigue stress's effect on life is an exponential factor, we know that doubling the load may cut the life by a factor of 10. From what we have seen, this also applies to gear tooth deterioration but it may be conservative, i.e., doubling the load reduce the life by more than a factor of 10.

The best way we have found to monitor the actual contact is to look at the gears in operation using a strobe light and an infrared scanner, watching the tooth appearance and the temperature distribution both across the teeth and around the entire gear. On the large open gear sets we try to keep the variation to less than 6°C (≈10°F). This inspection is easy to do on an open gear set but almost impossible in an enclosed unit. However, we have run enclosed units, removed the bolts on the inspection plate until only two or so are left, and then shut the unit down and quickly opened the cover, taking the temperature readings as soon as the unit stops moving.

(When we started the Reliability Program at the plant where I worked, we had no idea of the value of good alignment. As our alignment techniques improved, the life of our gears increased amazingly and our gear cost per ton of product decreased measurably.)

BACKLASH

One of the definitions earlier in this chapter states that *backlash* is "the effective clearance between the mating teeth." Measuring the backlash between a set of gears is fairly simple. One of the gears is locked in position so it can't move, then the other gear is rotated back and forth, and the clearance between the two meshing teeth is measured with a dial indicator or feeler gauge. With case-hardened gears, the backlash typically changes little over the life of the gears because there is usually little or no measurable wear, but with through-hardened gears, the backlash continually increases.

If the gears always drive in one direction, this increase is generally not very important because the load is continuous and the increase is really a measure of the tooth wear. With reversing gear sets, wear and increasing backlash can become a source of impact loading and must be evaluated. With the gears shown in Photo 11.7, the increase in backlash had reached more than 5 mm (0.2 in.), but that wasn't a problem because the motor controller started the system slowly and there was very little impact force involved.

PHOTO 11.7 Normal dedendum wear of this mine host gear was regularly tracked using a gear tooth micorometer.

DESIGN LIFE AND DETERIORATION MECHANISMS

Understanding the design life of a gear set is critical to understanding the true causes of a failure. The AGMA standards for reducer design call for an L_{10} life of 5000 h. This is only 5/8 of a year and the normal maintenance organization would expect any reducer to last far longer than that. The way the designer compensates for what the customer actually wants is to change the service factor, i.e., the input power rating of the reducer. In the 1960s it wasn't unusual to have a 3+ SF rating on a reducer, and that unit would typically last for ten plus years before any maintenance was needed, despite the fact that it was running in wet, dirty conditions with lubricants changed infrequently.

But competitive pressures have reached the reducer industry, and we have recently seen several 3- and 4-year-old failed reducers that have been bought with SFs in the area of 1.5. In those cases the user got exactly what they paid for.

The AGMA categorizes gear failures into categories of wear, surface fatigue, plastic flow, breakage, and associated gear failures. (We approach the failure analysis procedure from a different perspective but think their literature on gear failures is an excellent source of information.) Looking at their classifications of gear failures, they include:

1. *Wear* — The term "wear" covers a huge range of deterioration rates from mild polishing of gears that may last more than a hundred years to adhesive and abrasive wear that can render the gears essentially useless in a few minutes. For through-hardened gears, most wear is predictable and can be monitored. For case-hardened gears, any wear more substantial than polishing is usually a sign of impending failure.
2. *Surface fatigue* — This includes pitting and spalling and again, from a practical standpoint, surface deterioration of case-hardened gears is worrisome.
3. *Plastic flow* — With surface-hardened gears, this includes rippling, a relatively slight deformation of the gear surface. Through-hardened gears frequently suffer from rolling and peening that leave the surface profile badly damaged.
4. *Fracture* — In addition to the classical fatigue and overload fracture mechanisms we've covered in earlier chapters, case-hardened gears sometimes suffer from case crushing where the contact loads are so great that the underlying structure deforms, cracking the hardened case.

PHOTO 11.8 Wear of a through-hardened pinion with less than 50% of the tooth remaining.

THROUGH-HARDENED GEAR DETERIORATION MECHANISMS

PITTING AND WEAR

Gear deterioration can be very confusing and complex. Through-hardened gears are frequently not very hard and not very precise, and they tend to pit and wear in a relatively predictable manner. However, with case-hardened gears, any surface distress is cause for concern and pitting is often a sign of an impending disaster. We suspect that this huge difference in operating appearances leads to many misunderstandings about through-hardened tooth wear and probably results in more unneeded gear replacements than anything else.

A good side of through-hardened tooth deterioration is that it results in relatively predictable degradation, and it isn't uncommon to see through-hardened teeth wear substantially before there should be any concern about failure. We have seen several smoothly operating older reducers where less than 20 % of the original tooth thickness remained and yet they were still doing their jobs. In many through-hardened applications the condemning thickness is on the order of 60% of the original tooth thickness and sometimes it is even less. The gear shown in Photo 11.8 had run for 11 years in a large printing press. It was removed from service but was still functioning well with more than 50% of the tooth gone. (An inspection of the photo also finds that this relatively soft through-hardened gear was poorly aligned.)

There are a number of through-hardened tooth deterioration mechanisms including pitting, adhesive wear, rolling, and peening, but most gear wear starts with pitting and, with the combination of loads, materials, and lubricant film, it is frequently unavoidable. Figure 11.9 shows three general classifications for pitting and their common locations.

The mechanisms are:

1. *Corrective pitting* — As two gears mesh together, the high spots rub against each other and Hertzian fatigue causes small spalls. As time goes on, more and more of the tooth surfaces contact each other and the wear pattern becomes wider and wider. These pits are relatively small, and generally shouldn't exceed 1.5 mm ($\approx$ 0.06 in.) in diameter. They support the development of a lubricant film and are in the dedendum of the teeth. Corrective pitting starts rapidly but as the contact area increases the wear rate decreases, eventually reaching a low but continuous level. (There really is a difference between wearing in and wearing out!)

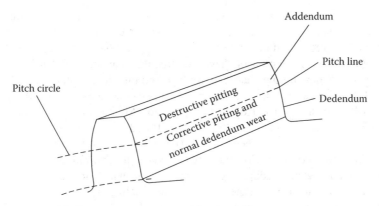

FIGURE 11.9 Common pitting locations on a through-hardened gear.

2. *Normal dedendum wear* — This occurs in the dedendum of the teeth as a long-term function of the tensile fatigue stresses. It generally shows up in the dedendum of the teeth as the gears age and the deterioration rate is very slow and readily monitored.
3. *Destructive pitting* — These pits are large and irregular and don't support the lubricant film. They begin in the dedendum but rapidly progress into the addendum with the resultant tooth wear and noise.

(Texts on gears usually speak of corrective pitting being 1 mm in diameter or so, but the dividing line between acceptable corrective pitting and worrisome destructive pitting is not very clear, and from a practical standpoint, the pitting to be concerned about varies with the speed of the gears and the lubricant. The pits shown in Photo 11.10 are much larger than those shown in Photo 11.8 (and more than 3 mm [1/8 in.] wide in some cases) yet both gears will last more than 20 years. The gears in Photo 11.8 are in an enclosed reducer and fed with a pressurized and filtered spray using an ISO 320 gear oil. Those in Photo 11.10 use a heavy open-gear grease, and the pits will grow smaller once the pattern extends all the way across the tooth.)

One of the difficulties in assessing through-hardened tooth wear occurs shortly after startup, when the teeth begin pitting, and it is initially difficult to tell the difference between corrective and destructive wear. Figure 11.10 shows a chart of the pitting rates in a typical gear installation, but sometimes monitoring of pitting can be difficult. In the past, carbon paper was run between the

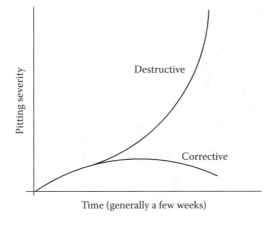

FIGURE 11.10 Tracking the difference between destructive and corrective pitting over time.

teeth to measure tooth mesh changes, but we generally use a combination of oil sampling, digital photography, and measurements with a gear tooth micrometer. The figure shows the difference showing up over a few weeks, and in a typical industrial application, if the wear rate doesn't start to decrease within 4 to 6 weeks, you probably have a destructive wear situation that will only get worse.

An example of both how pitting changes with load and the predictable nature of pitting can be seen in Photo 11.8. This is a mine hoist gear that lifts a personnel elevator over 0.6 km (≈2000 ft), and it was originally driven by a 1300 kw (1000-hp) motor. After about 15 years of operation, they decided to increase the motor to 2000 kw (1500 hp). Based on the original pitch line thickness, the normal dedendum wear rate had been about 0.05 mm/year (0.002 in./year). For many years after the motor was replaced, we periodically measured the tooth width just below the pitch line using a gear tooth micrometer. Readings were taken once or twice per year in six positions across the tooth and at three positions around the gear. Then the readings were logged and the tooth wear was charted at approximately 0.2 mm/year (0.008 in./year). With a calculated condemning thickness well in excess of what would ordinarily be used in an industrial application, this not only gave the mine a predictive maintenance tool that they could use to justify replacement gears, it also prevented an unnecessary early expenditure. (In this example, increasing the available motor power by 50% increased the wear rate by about a factor of 4.)

The gear tooth micrometer is essentially two sets of vernier calipers that are joined together and is shown in Photo 11.9. For field measurements, we normally set the depth about 20% below the pitch line, then measure and chart the tooth width. (One caution about using the micrometer is that you have to be sure there isn't a lip from plastic deformation on the top land of the tooth.)

Photo 11.10 shows a through-hardened gear that has been removed from a large piece of mining equipment. The teeth show some dedendum wear and some of the pits are larger than we would like to see, but it had been in run for 8 years and a decision was made to change it despite the fact that it has many years left on it. Looking at this gear, we can see that the pattern has not extended all the way across the teeth and the tooth profile is still in very good condition. This is an ideal application for periodic inspection with the gear tooth micrometer, and this gear should be carefully stored and preserved as a viable spare.

We shouldn't leave this example without at least mentioning the subject of corrosion of gear teeth. In the chapter on fatigue, we discussed at length how corrosion decreases the fatigue strength of metals. With through-hardened teeth that are relatively soft, less than about HBN 180, corrosion will basically accelerate the pitting action and will result in somewhat shortened gear life. As the teeth become harder, the effect becomes much more serious. We know from numerous studies that

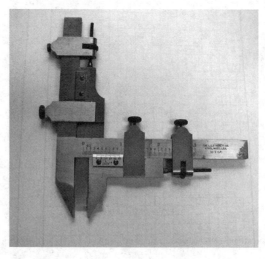

PHOTO 11.9 A gear tooth micrometer.

PHOTO 11.10 Corrective pitting of a through-hardened dragline swing gear.

atmospheric corrosion can cause the fracture of highly stressed materials with hardnesses as low as HRC 40. Our rule of thumb is that gears in storage should always be protected from corrosion. In addition, on a case-hardened gear, any visible rust is proof of damage that will reduce the fatigue life and visible heavy rust has probably already resulted in microcracking of the case.

Plastic Deformation

Investigating other common through-hardened tooth deterioration mechanisms, we see that rolling and peening involving plastic deformation of the teeth. In both situations the loads are heavy enough to physically deform the teeth. *Rolling* involves contact stresses large enough to physically move the metal up over the top or around the side of the tooth, resulting in a sharp ridge over the top land of the tooth or a mushroomed appearance on the side. *Peening* is the hammering of the mating teeth resulting in a deformed (peened) tooth profile. In both cases, either the lubrication has been inadequate or the loads have been excessive, and with continued operation, the gears will grow progressively noisier and rougher.

Looking at the two bevel gears in Photo 11.11, the one on the right is like new but the one to the left has lost a significant portion of the tooth width. Wear has thinned the teeth and rolling has moved the metal around the edges of the teeth, but in a heavily loaded low-speed application the unit still has many more productive hours left in it. (Looking at the rolling damage, this is a reversing

PHOTO 11.11 A through-hardened steel mill gear with heavily worn teeth that show rolling damage.

PHOTO 11.12 Scuffing (adhesive wear) of a through-hardened tooth.

gear, and as long as the impact loading from the increased backlash isn't excessive, we expect this gear would have traveled about 2/3 of the way through its usable life.)

ADHESIVE WEAR

There are other terms commonly used to describe adhesive wear with scuffing, scoring, galling, and seizing being the most frequently heard. Adhesive wear involves breakdown of the lubricant film and momentary welding of the teeth, followed by tearing a particle out of one of the teeth. It is most often seen at the tips and the roots because those are the areas where it is most difficult to develop an effective lubricant film. Photo 11.12 shows a close-up of a worn tooth and the radial marks from poorly lubricated contact with the opposing teeth are plainly visible. The wear rate decreases as the contact surface approaches the pitch line. At the pitch line there is pure rolling, and both there and above it, there is a greatly improved lubricant film and decreased wear rate. (If the contact stress had been lighter, this "scuffing/scoring" would likely be replaced by polishing and the metal removal rate would have dropped substantially.) This is a graphic example of the effect of the combined rolling and sliding action.

CASE-HARDENED GEAR DETERIORATION MECHANISMS

Earlier we said that case-hardened gears should show essentially no wear or deterioration, and our experience has shown that for most of these gears this is true. However, there are occasional trouble areas and warning signs that should be recognized.

MICROPITTING

Micropitting occurs when a surface is heavily loaded and has the appearance of a lightly sandblasted surface. (Sometimes it is called "frosting," but we feel micropitting is a more accurate descriptive term.) In some instances micropitting is of absolutely no concern, but in others it is a danger warning. Photo 11.13 shows a heavily loaded 1350-kW (1000-hp) reversing drive from a steel mill with obvious diagonal patterns across both sets of teeth. These patterns are a series of very fine pits and, in this case, the micropitting:

- Occurred because of the slight irregularities on the machined surface
- Is absolutely nothing to be concerned about
- Shows that the gears are heavily loaded
- Shows that the alignment is excellent

PHOTO 11.13 The harmless micropitting on these teeth show up a bands where the machining has left minute ridges.

PHOTO 11.14 This micropitting is a symptom of serious misalignment.

In many situations however micropitting is cause for concern as is shown in the two examples below. Photo 11.14 shows a 3.2 cm (1.25 in.) wide centrifugal air compressor drive gear with micropitting on the left side of the teeth. Photo 11.15 is of a gear from a small generator drive that shows micropitting on the right side. In both applications the fact that the attack is concentrated on one side and is relatively obvious indicates serious alignment problems then, looking at Photo 11.15, there is a pit on the lowest tooth in the photo.

One of the design requirements is that surface-hardened gears must be well aligned. The elasticity of the teeth will correct for minor misalignment and micropitting along one side of a gear is a warning that there are both very high local loads and excessive misalignment.

RIPPLING

The appearance of *rippling* on a case-hardened gear falls in the same basic analytical category as micropitting. It tells that the gear has been very heavily loaded and there is some plastic deformation of the surface. Again, as with micropitting:

1. If the rippling is consistent across the tooth, it isn't as worrisome as if it is on one side of the tooth. It is a sign that the gear should be closely monitored but not an indication of an immediate concern.
2. If the rippling is on one side of the tooth, it is a tell-tale sign that there is misalignment, a warning that a failure may occur in the future, and a call for regular inspections to ensure that there isn't further deterioration.

PHOTO 11.15 Micropitting from misalignment plus a pit, in this case hardened gear.

PITTING

With the knowledge that micropitting shows there are heavy loads, any macropitting is source of serious concern. Having said that, we should emphasize that pitting can't automatically be considered a warning of an impending disaster, but it does have to be carefully analyzed along the lines of any other failure.

Our logic about being concerned about pitting of surface-hardened gears goes back to the tensile strength of the teeth. When the case is HRC 55, its tensile strength is approximately 1.9 GPa (275,000 psi). If the core is HRC 35, its tensile strength is less than 2/3 that of the case. If the load is heavy enough to break the case, how long would a core last that is only 60% as strong?

One of the problems with this approach is that to some degree it is comparing apples with oranges. The case damage occurs because of Hertzian fatigue, with complications from sliding, whereas the tooth breakage results from cantilever loading. Nevertheless, probably 99% of the surface hardened teeth we've seen that have failed from low or high-cycle fatigue have had pitting problems. In the analysis of pitted, surface-hardened teeth, look carefully at the shape of the pits. Oblong pits, with their long axis along the length of the tooth, and relatively deep sharp bottoms are serious stress risers. On the other hand, shallow, round pits with round bottoms are not nearly as problematical.

In Photo 11.16 we see most of the active profile a case-hardened tooth. The bottom of the photo ends at the root radius and the pits are starting to become elongated and link together, a source of concern. In addition, careful inspection shows some deterioration on the right side of the tooth and faint indications of rippling on the left side. Any recommendation on the future of the gear would depend on how long it had taken to reach this state. However, our opinion is that we would be very reluctant to run it in any sort of a critical drive.

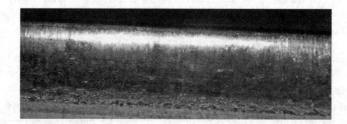

PHOTO 11.16 This case-hardened tooth shows worrisome pitting along with rippling.

PHOTO 11.17 The center of this photo shows a spall that is ½ the height of the tooth — a sure sign of manufacturing problems.

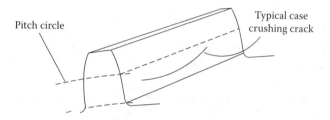

FIGURE 11.11 A sketch of case crushing with the crack running along the length of the active profile.

SPALLING

Small pits are an indication of high localized stresses and appear on heavily loaded gears. Large spalls are an indication of metallurgical problems in the manufacture of the gear. Photo 11.17 is a bit of a difficult photo to view because it was taken through an expanded metal guard. However, in the center of the photo is a spall that covers more than half of the active flank of a 3.8-cm (1.5-in.) tooth. The fact that there is this huge area of separation between the case and the core is almost always evidence of a heat-treating problem.

CASE CRUSHING

Figure 11.11 is a sketch showing case crushing on a surface-hardened gear. It happens when the load on the tooth is so great that it causes plastic deformation of the softer core, but the case is harder and less ductile than the core, and can't tolerate the deformation and fractures. The sketch shows the cracking in the dedendum, but it can occur anywhere on the tooth and is characterized by horizontal cracks.

ANALYZING GEAR FAILURES

Approaching a gear failure, the goal is to understand why it happened and the analysis should be done in the same basic manner as any other failure analysis. The difficulty with a gear failure analysis is that they are very complex and it is not easy to separate some of the mechanisms. However, start by categorizing the type of failure as shown in Figure 11.12 and then look at the various features, i.e., the last two tiers of this chart.

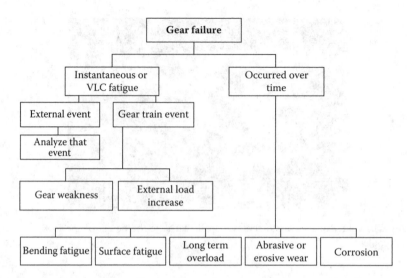

FIGURE 11.12 A basic gear failure analysis tree.

The characteristics that affect gear life and some questions that should be used as a starting point for the analysis are:

1. *Load* — Are either the Hertzian fatigue or cantilever loads excessive?
2. *Load distribution* (or *alignment*) — Is it uniform around the gear and across the teeth?
3. *Temperature and lubricant film thickness* — How do the operating temperature and the lubricant package compare with the original specifications?
4. *Gear tooth geometry and finish* — How does the current tooth profile (actual tooth shape) compare with the original gear specifications? Was it in storage and allowed to rust?

Some notes on Figure 11.12 are:

1. We've combined instantaneous and very-low-cycle (VLC) failures in one group because, even though gears are very strong, it isn't unusual for an external event to occur that causes the gear to fail very rapidly. These failures typically have the same fracture appearances as any other brittle overload with comparable material.
2. In following that path on the right, those failures that don't occur in a short time, we typically use slightly different thought processes for surface-hardened gears vs. through-hardened gears but the differences aren't substantial, and the following notes are written as though the process is identical.

Bending fatigue cracking indicates the cantilever strength of the tooth has been exceeded over some time. The questions are designed to help find why the load was excessive.

1. For all gears
   A. Is it a reversing application?
      a. If so, what sort of impact occurs when the load is reversed?
      b. Was there wear on the teeth? How much?
   B. Where does the fatigue crack start? Is it at the center of the teeth (showing good alignment) or far toward one edge?
   C. Are there fatigue cracks on many teeth, indicating a very high overload, or only on one or two teeth, showing the loading was only slightly above the fatigue strength?

(In answering this question, consideration has to be given to the time since the first fatigue crack originated. There may be many fatigue cracks because the gear has been forced to run for a long time since the cracking started.)

D. Where is the contact pattern? Has it changed recently? Is there any evidence of contact on the unloaded side of the teeth? Is there obvious wear of the teeth?

a. For case-hardened teeth, variations in the wear pattern may be difficult to see and it may be necessary to slowly rotate the gear with a bright light on the teeth.

b. For through-hardened teeth:

1. Some wear is expected but it may have gone too far. Determine the condemning thickness.

2. Is the wear uniform across the teeth and around the gear? If it isn't, it shows the gears are either misaligned or mismachined.

E. What was the original material specification? What is the hardness specification and what is the actual tooth hardness?

2. For case-hardened teeth

a. Is there any evidence of micropitting? Is the pattern skewed, indicating eccentric loads or misalignment?

b. Is there any evidence of plastic deformation, i.e., a thin ridge along the top of the tooth?

c. Did the cracks originate at pits, evidence of very high loads?

3. Was the lubrication as specified? What load was the gear originally designed to carry? What was the service factor? How does the load vary with time? Have the loading variations been adequately verified?

4. Was the gear ever allowed to rust and corrode, greatly reducing the fatigue strength?

Surface Fatigue — Hertzian fatigue spalling, pitting, and micropitting happen when the surface loads are excessive and a frequent contributor is that the lubricant film is inadequate.

1. Where does the spalling or micropitting start? Is it at the center of the teeth (showing good alignment) or far toward one edge?

2. Look at the mating gear. What is the tooth profile like? Is it in good condition?

3. The correct lubricant film should prevent any pitting. What has the lubrication history been and how are the lubricants applied? Are EP lubricants used? Do the change/application intervals recognize the possible deterioration of the EP lubes? How does the lubricant operating temperature compare with the design operation temperature? What have the lube samples shown?

4. Is the surface fatigue pattern uniform around the gear, indicating good gear geometry, or is it inconsistent, indicating machining errors?

a. For through-hardened teeth, is the wear uniform across the teeth and around the gear? (If it isn't, it shows the gears are either misaligned or mismachined.)

b. With case-hardened teeth:

i. Some light and uniform micropitting is not of concern. But, with the alignment needs of hardened materials, if the pattern is skewed or erratic, it indicates a potential problem.

ii. Eccentric micropitting, on one side of a tooth, is an indication of abnormally high surface stresses and alignment problems.

iii. The presence of pitting on through-hardened gears is another result of abnormally high surface stresses and is a great concern.

iv. An isolated large spall (or spalls), greater than 6-mm (¼-in.) diameter is almost always an indication of manufacturing problems.

v. Was the gear ever allowed to rust and corrode, greatly reducing the fatigue strength?

Long-term overload — This primarily occurs on through-hardened teeth and usually results in slow and predictable wear, a result of heavy loads with weak lubrication.

1. Are the teeth worn excessively, spalled, or broken?
2. Is the damage uniform around the gear?
3. Was the loading across the teeth correct?
4. Is there any micropitting or pitting?
    * Is the pitting in the addendum, dedendum, or both?
    * How large are the pits?

FAILURE EXAMPLES

INTERMEDIATE PINION

Photo 11.18 shows a case-hardened intermediate pinion on a reducer that had run for about a year. This is a three-stage reducer was removed from service because of a vibration analysis indicating a problem. It had never been apart and the loads and speeds were well within the design specifications.

Inspection

On inspection, we found:

1. Another spall about 120° around the case-hardened intermediate pinion from where Photo 11.18 was taken.
2. The lubricant tests indicated the lubricant was as specified by the manufacturer.
3. Other than the two spalls there was no evidence of other damage and no measurable misalignment.

Conclusions and Recommendations

There is nothing the user can do to cause or avoid this type of spalling in an intermediate gear set. We suspected a metallurgical problem and recommended a detailed metallurgical analysis to determine the true physical roots. This found there was a processing problem.

PHOTO 11.18 Spalling of an intermediate spiral bevel pinion — the result of a manufacturing problem.

PHOTO 11.19 This case-hardened pinion failed from fatigue but was poorly aligned.

BROKEN GEAR TOOTH

This was a single tooth from a large, slow gear in a materials processing plant. It had run for about 6 months at 5 r/min when the tooth failed.

Inspection

Inspection of the fracture face found, as shown in Photo 11.19:

1. The tooth has a relatively thick case tempered to about HRC 52. (We have blackened the case surface in the bottom center of the photo.)
2. The tooth failed from fatigue, but knowing that the steel has a relatively fine-grain structure, the failure was very rapid.
3. The load was poorly centered, and the crack started far to the right side of the tooth and the left half of the tooth was the instantaneous zone.

Inspection of the contact surface of the tooth found some pitting that was also centered far to the right side of the tooth.

Conclusions and Recommendations

The gears were seriously misaligned and our recommendation was to correct the misalignment. (This is another example of the sensitivity of case hardened gears to misalignment.)

LARGE PUMP DRIVE GEAR

Photo 11.20 is of a 4700 kw (3500-hp) gear out of a pump drive that has two 2350 kw (1750-hp), 1800-r/min motors driving pinions that meshed with this gear. The output speed was about 400 r/min. This is a flame-hardened tooth with a proprietary contour and the heat-treated profile is

PHOTO 11.20 The river marks in the failure face show this cracking progressed from left to right.

PHOTO 11.21 The large spalls on this pinion lead one to suspect a metallurgical problem.

readily visible on the side of the teeth. The drive had been in operation for more than a year, and there had been several comments from technical and maintenance personnel about how noisy it was.

Inspection

Inspection of the fracture face shows that the gear is driven to the left, but the river marks tell us that the crack growth is from left to right, against the drive forces. An inspection of the tooth surfaces shows there has been repeated contact on the "unloaded" side of the teeth.

Conclusions and Recommendations

There was a torsional resonance in the gear teeth, i.e., the teeth were being struck or driven at a frequency that would excite their natural frequencies (confirmed by vibration analysis). The reversing forces put the weakest section of the case into tension, causing a crack that then propagated in the much weaker core section. As shown earlier in the chapter, for there to be contact on the unloaded side of the tooth, the tooth driving forces had to be more than twice the design forces. Changing the natural frequency of the gear and shaft assembly would be extremely expensive and we recommended they retire the reducer.

HAUL TRUCK PINION

Photo 11.21 is of a pinion from the rear wheels of a >300-ton haul truck. This drive axle drives the truck uphill and is loaded on the other side by the braking system in the downhill run; it had been in operation about 6 months.

Inspection

Inspection found this tooth shows two large spalls and another tooth (not in the photo) showed a single large spall near the left end of the tooth. There is no visible contact pattern in the photo, and the actual contact pattern looked good on the more heavily loaded side.

Conclusions and Recommendations

The large size and unusual locations of the spalls lead us to suspect a materials problem possibly complicated by misalignment. We recommend:

- Carefully reinspecting this gear using wet fluorescent magnetic particle testing to determine if there is any more cracking. (This may show a more definite pattern.)
- Having a metallurgist section the tooth and analyze the metallurgy, and the heat treatment.

Paper Machine Reducer Gear

This reducer had been on the paper machine for over 10 years, but the gears had been replaced about 4 years before the failure. Shortly before the reducer failed, the cover had come off the driven roll, requiring a shutdown. When the machine was restarted they noticed the reducer was very noisy and they replaced it.

Inspection

Inspection found three broken teeth on this through-hardened gear. The oldest, based on the appearance of the fatigue crack surface, is shown in Photo 11.22, and the other broken teeth were spaced around the gear. Further inspection of the gear showed that it had been misaligned and wearing irregularly as shown in Photo 11.23. There were some larger pits, but most of the pitting was relatively fine and was a combination of corrective pitting and normal dedendum wear. The historical data on the reducer lubricant appeared reasonable and hardness testing of the gear showed it to be HRC 35. Gear tooth micrometer measurements showed they had worn by about 8%. (There was no drive motor load data to look at.)

Conclusions and Recommendations

The wear pattern seen in the second photo shows these through-hardened gears were obviously misaligned but they should not have failed with only 8% of the tooth missing. The surface of the

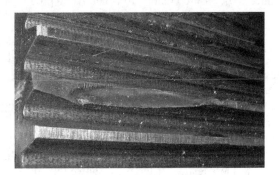

PHOTO 11.22 This was the first of three broken teeth on the gear.

PHOTO 11.23 Notice the contact pattern disappears on the right side of the teeth.

PHOTO 11.24 The varying pitting on these teeth show there are machining inconsistencies.

fatigue crack indicated that the gears had only run only a short time between initiation of the cracking and the point when the tooth failed and caused other damage, but the fracture shows there was a very heavy load on the tooth. Based on this, our conclusion was that the failure of the cover increased the reducer load for a short time and resulted in the gear failure. We also recommended they contact the reducer manufacturer to discuss the internal misalignment and possible correction.

KILN REDUCER INTERMEDIATE GEAR

This kiln had been in operation for over 20 years when the management started considering increasing the motor horsepower by 10% to allow the kiln to rotate faster. The reducer gears had not been a problem, and they were not certain they would increase the kiln speed. The primary interest was that they wanted to be assured that there would be no problems over the next 10 years.

Inspection

This was a three stage reducer and the first five gears looked like they did on the day they were manufactured. The tooth profiles showed no wear and the contact patterns showed light polishing all the way across the gears and were consistent around all five gears.

The low-speed-output gear had 63 teeth and typically rotated at 20 to 24 r/min, depending on the kiln speed. Inspection of this gear showed irregular pitting on several of the teeth as shown in Photo 11.24. (That is an oil spray pipe across the center of the photo.) On some of the teeth there was no pitting, but on others it extended more than half way across the tooth. There was no evidence of plastic deformation and hardness testing of the teeth showed them to be HRC 31.

Conclusions and Recommendations

The low-speed gear shows corrective pitting, a result of minor machining errors while the gears were being manufactured. With this little wear over the last 20 years, we would expect the reducer to last more than another 20 years even if the horsepower were to be increased by 10%.

INBOARD–OUTBOARD PROPELLER DRIVE GEAR

This wasn't the subject of a lengthy investigation. It is the gear off an outboard drive on a boat that is frequently used by a friend to go out to sea for sport fishing. Eventually, it failed and they asked us to look at the gear. There were other problems, but two interesting points can be seen looking at Photo 11.25:

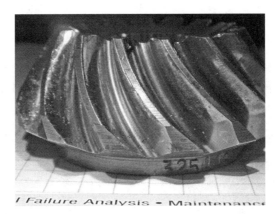

PHOTO 11.25 This spiral bevel pinion shows numerous symptoms of heavy loads and poor lubrication.

1. On the teeth to the left, there is plastic deformation of the contact surface.
2. In the center of the photo, there is a small but obvious ridge of plastic deformation along the top lands of the teeth.

We asked if they regularly checked and changed the lower unit lubricant. The answer was a sheepish "no," but they promised to be more conscientious in the future.

SUMMARY

Follow the outline and look at every symptom. The gears will tell you why they failed.

12 Fastener and Bolted Joint Failures

When we started conducting training programs at the big, wet, dirty caustic chemical plant where I worked, the first workforce classes involved fasteners. We had some truly impressive disasters with a variety of bolt failures; therefore, the emphasis in those early classes was on preventing bolt breakage. Eventually, we sat back and did some more detailed analyses of our situation and realized that the major physical cause of most of our leakage problems was poor bolting practices. Since that time, we've worked with hundreds of plants across North America, and invariably find that a major (if not the major) maintenance cost is repairing leaks. Flange leaks don't just happen, and those steam and process leaks that required expensive cold shutdowns to repair are almost always unintentionally caused by the people that put them together.

In this chapter, we hope to:

1. Provide data on the common fastener materials available.
2. Help you to understand how bolts work and how they fail.
3. Explain the importance of uniform tightening procedures, emphasizing that proper flange assembly requires intelligence, care, and skill.
4. Help you to diagnose fastener and bolted joint failures.

HOW BOLTS WORK

The idea of bolt elasticity is a critical one, and a bolt acts a bit like a stiff rubber band as it is being tightened. This stretching and the resultant clamping force then hold the parts together. In Figure 12.1, the sketch of a typical bolt includes the grip length, which is important because it is the stretch in this area that provides the elasticity that does the clamping, the property critical to maintaining effective bolting performance.

With equal loads, all grades of steel stretch the same amount, and the benefit of using a stronger bolt is that the elastic range of the steel is greater, and the bolt can stretch more before it plastically deforms. Figure 12.2 gives some practical data on different grades of steel and the corresponding stretch. Remember though, there is no such thing as a free lunch. The trade-off for increased strength is increased susceptibility to fracture from corrosion and reduced toughness.

With a conventional nut and bolt assembly, as it is tightened and the assembly develops this clamping force, some of the energy is absorbed by the stretching whereas the rest of it has to overcome friction, either between the rotating piece and the mating threads, or between the rotating piece and its contact surface. Therefore, with better lubrication and less friction, more of the tightening torque can go into developing the clamping force. There have been many studies of this area, and the current thought is that about 40% of the turning effort goes into overcoming thread friction, about 42% goes into overcoming friction between the rotating member face and the piece being clamped, and the remainder goes into stretching the bolt.

Looking at the action of this clamping force, one sees that it is absolutely critical to effective performance. A good example is a bus or truck wheel. The wheel bolt's function is to tightly clamp the wheel against the hub so the friction between the two members can drive the wheel and, if the bolts can't provide enough clamping force, the bolts see a fatigue load and rapidly fail.

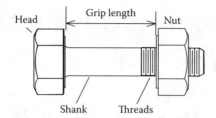

FIGURE 12.1 A typical bolt with the terminology used in this book.

Approximate tensile strength (metric/US)	Typical grades	Allowable stretch (mm per mm of grip) (in. per inch of grip)
400 to 500 MPa/80,000 psi	M – 4.6, 4.8, 5.8 US – A 307, SAE Gr. 1 & 2	0.001
800 MPa/105,000 to 125,000 psi	M – 8.8, 9.8 US – A 325 & 449 SAE Gr. 5	0.002
1000 MPa/150,000 psi	M – 10.9 US – A 490, SAE Gr. 8	0.003

FIGURE 12.2 This shows how the tensile strength affects the allowable enlongation of the bolt.

We will discuss bolting practices later, but a wheel hub on a car is also a great example of the need for good tightening procedures. If the lug bolts aren't tightened adequately, they will fail from fatigue and the wheel will fall off. On the other hand, if they are tightened either too much or nonuniformly, they will cause distortion of the wheel hub, warp the brake disk or drum, and result in all sorts of problems.

Three important points to understand are:

1. The clamping force must be greater than the separating force if fastener fatigue is to be avoided.
2. On a gasketed assembly, uniform clamping is critical to minimize warpage and distortion and give good performance.
3. Overtightening can cause not only component distortion but also thread seizure and possible fracture.

Pursuing this, several years ago the Bolting Technology Council (This was a technical group that has since been absorbed by and whose work has been continued by the ASTM.) hired Andre Bazergui of L'Ecole Polytechnique of Montreal to analyze those things that could affect torque vs. tightness in a bolt. Dr. Bazergui did an excellent job, with a detailed study looking at an exhaustive variety of possible conditions and modifiers and reported that the most important properties, in order of severity, are:

1. Proper installation (install the bolt squarely)
2. Thread lubricants and plating
3. Materials
4. Installation practices
5. Everything else, i.e., corrosion, surface smoothness, tightening method, number of threads engaged, hand or power tightened, etc.

From this study, we can see the importance of consistent bolting practices, and we will return to that later in the chapter.

Bolt Standards

There are numerous groups around the world that have standards concerning bolts and, combining the different bolt strength levels and materials, there are hundreds of possible material combinations. In turn, those bolts are made by thousands of different companies, and one of the questions a failure analyst frequently hears is "How do you know how strong the bolt is or who made it?" The answer can be found by looking at the bolt head and the markings.

According to U.S. law, the Fastener Act of 1991, and ISO standards, unhardened low-carbon steels, the weakest of steel bolting materials, don't require any head markings. However, all heat-treated general-purpose bolts and studs must have the grade and manufacturer's identification clearly marked on their heads, and Figure 12.3 shows a list of the common U.S. threaded bolt specifications and their identification symbols. (In addition, ISO standards require the manufacturer's identification on all strain-hardened bolts.)

> How do you find out who made that bolt? *Fastener Technology International*, a magazine from Cleveland, OH, maintains a registry of manufacturers and their symbols. To determine who made a fastener, you can buy a copy of their roster of trademarks and search. Some are easily recognized abbreviations such as RBW (for Russell, Burdsall, and Ward), whereas others are relatively hard-to-interpret symbols, but the registry contains essentialy all the major suppliers from around the world.

General-purpose metric bolts can easily be identified because they have their strength markings on the heads, always using a decimal point between the tensile strength value and the proof test multiplier. Photo 12.1 shows a metric bolt head, and Figure 12.4 shows a chart of the common metric fastener designations along with the comparable SAE and ASTM standards. The first number on the head shows the tensile strength in hundreds of megapascals while the number after the decimal point. is the proof test multiplier. In the example in the photo, the 8.8 shows a bolt in which the tensile strength is approximately 800 MPa, the proof test multiplier is 0.8, and the resulting proof strength is approximately 640 MPa. In this example, the "SBE" on the bolt head is the manufacturer's trademark.

We should be very careful to note that all U.S. thread steel socket-head cap screws (Allen head cap screws) are quenched and tempered medium-carbon alloy steels with a minimum tensile strength of 150,000 psi (1,040 MPa). This is the same material specification are used for an SAE Grade 8 bolt but it only applies to steel U.S. thread socket-head cap screws. It doesn't apply to stainless steel or other alloys, and metric socket-head cap screws can be found in all seven classes listed in Figure 12.4.

Nut Standards

Just as there is a variety of bolt strengths, there is a variety of nut strengths; however, because of the way the stress is applied, nuts are inherently stronger than the bolts or studs they are attached to. Visualizing a cross section of the assembly shown in Figure 12.1, the shank of the bolt and the bolt threads are in tension, whereas the nut body and threads are in compression.

It is extremely uncommon to see a properly matched nut fail. They cannot fail from fatigue (because the threads and the body are in compression), and the actual mechanism is almost always an overload failure with dilation from overstress, followed by the threads stripping. There are numerous steel nut standards but a good rule of thumb to follow is shown in Figure 12.5.

What is confusing about nuts is how they are identified. Metric nuts, as with metric bolts, are marked with the strength class, but U.S. thread nuts have several different systems as seen in Figure 12.6.

Grade designation (ASTM & SAE)	Material and condition	Tensile strength (ksi, min)	Proof load (ksi, min)	Size range (inches)	Head identification
General usage bolts					
A 307 Gr A	Low/medium carbon steel not heat treated	60	na	1/2–4	None required- may have trademark
A 307 Gr B	same	60 min 100 max	na	1/4–4	None required may have trademark
SAE Gr. 1	same	60 min	36	1/4–1.5	None required may have trademark
A 325	Low/medium carbon steel, Q & T	120 / 105	85 / 74	1/2–1 / 1.1–1.5	A 325 Trademark req'd
SAE Gr. 5	Medium carbon steel, Q & T	120	85	1/4–1.5	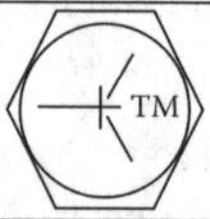 (hexagon mark) TM
A 354 Grade BC	Medium carbon alloy steel, Q&T	125 / 115	105 / 95	1/4–2.5 / 2.6–4	BC Trademark req'd
A 354 Grade BD	Medium carbon alloy steel, Q & T	150 / 140	120 / 105	1/4–2.5 / 2.6–4	BD Trademark req'd
A 449	Medium carbon steel, Q & T	120 / 105 / 90	85 / 74 / 55	1/4–1 / 1.1–1.5 / 1.6–3	A 449 Trademark req'd
SAE Gr. 8	Medium carbon alloy steel, Q & T	150 min 170 max	120	1/4–1.5	 (hexagon mark) TM
A 490 Type 1	Medium carbon steel, Q & T	150 min 170 max	120	1/2–1.5	A 490 Trademark req'd
A 490 Type 2	Low carbon alloy steel, Q & T	150 min 170 max	120	1/2–1	A 490 Trademark req'd
Low temperature					
A 320 (tested @−101°C)	Medium carbon alloy steel, Q & T	125	105	up to 2.5	L7 Trademark req'd
A 320 (tested @−73)	Medium carbon alloy steel, Q & T	100	80	up to 2.5	L7M Trademark req'd
High temperature					
A 437	Martensitic stainless, Q & T	145	105	All	B4B
A 437	Martensitic stainless, Q & T	115	85	All	B4C
A 437	Medium carbon alloy steel, Q & T	125	105	up to 2.5	B4D

FIGURE 12.3 Common inch-series fastener specifications.

MATERIALS

Everything we've written earlier has been about steel bolting materials but there are a huge variety of other bolting materials available. The charts in Figures 12.3 and 12.4 actually list about 10 different materials, but the current ASTM section on fastener standards cover more than 500 pages. Figure 12.7 shows some common ASTM standards for high-temperature alloy and stainless bolts and studs. In addition, there are other ASTM standards such as F 593 that covers general-purpose stainless fasteners and F 468 that covers nonferrous fasteners made of copper, nickel, aluminum, titanium, and their alloys. F 468 alone specifies 28 different alloys.

PHOTO 12.1 A general purpose metric fastener can be identified by the number (8.8) showing the strength level.

ISO class[1] identification (ASTM F 568)	Size range	Material[2]	Tensile strength (MPa)	Yield strength (MPa)[3]	Approximate SAE/ASTM equivalents[5]
4.6	M5–M100	Low or medium carbon steel	400	240	SAE Gr. 1 A 307 Gr A
4.8	M1.6–M16	Annealed low or medium carbon steel	420	340	SAE Gr. 1
5.8	M5–M24	Low or medium carbon steel, cold worked	520	420	SAE Gr. 2
8.8	M16–M72	Medium carbon steel, Q & T	830	660	SAE Gr. 5 A 449/A 325
9.8	M1.6–M16	Medium carbon or low carbon martensitic steel, Q & T	900	720	
10.9	M5–M100	Medium carbon steel, Q & T	1040	940	SAE Gr. 8 A 354 BD A 490
12.9	M1.6–M100	Alloy steel, Q & T	1220	1100	A 574

Notes:
1. Property class is the required identification on the fastener head or the stud body.
   All fasteners 5.8 and stronger must also be marked with the manufacturer's trademark.
2. Q & T = quenched and tempered
3. Proof strength as determined by the yield strength method.
4. SAE metric specifications are covered by Standard J1199
   Some metric bolts and studs may be marked with ASTM designations such as A 325M.
   The addition of the M indicates a metric fastener that meets the requirements of the ASTM standard.

FIGURE 12.4 Common metric fastener specifications.[4]

Therefore, there are not only a huge number of bolting materials but also a large number of plating treatments that can be used on those bolts. The reason for repeatedly mentioning this is that changing bolting materials has two significant effects:

Approximate bolt or stud tensile strength (metric/US)	Typical bolt grades	Required nut material
400 to 500 MPa/80,000 psi	M – 4.6, 4.8, 5.8 US – A 307, SAE Gr. 1 & 2	Any nut steel can be used but unhardened is preferred
800 MPa/105,000 to 125,000 psi	M–8.8, 9.8 US – A 325 & 449 SAE Gr. 5	Any nut steel can be used
1000 MPa/150,000 psi	M – 10.9 US – A 490, SAE Gr. 8	Nut should be made from steel that is comparable in strength with the bolt or stud

FIGURE 12.5 Guidelines for selecting the proper steel nut strength.

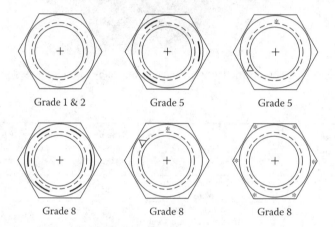

Grade 1 & 2   Grade 5   Grade 5

Grade 8   Grade 8   Grade 8

FIGURE 12.6 Identification of some US thread steel nuts.

1. As bolting materials get stronger, they become less able to absorb impact loads and more susceptible to corrosion damage and embrittlement.
2. Different materials and coatings have differing friction characteristics, and variations in the friction change the amount of torque that goes into stretching the bolt.

> If you cannot get a direct replacement material and have to change material specifications in a bolting application, either really know what you are doing or ask a professional for a recommendation in writing. Why do we say to get a recommendation in writing? As with everything else, specifications on nuts and bolts look fairly simple and straightforward from a distance. Up close, where you have to understand the effects of fatigue and impact loading and the possibility of corrosion and embrittlement failures, the job is much more complicated. By putting the recommendations in writing, the expert will carefully review what is actually happening.

FATIGUE DESIGN

Peterson's Stress Concentration Factors, edited by Pilkey, is a book that shows examples of components designed to reduce stress concentrations. One of the illustrations is of a bolt where the body is turned down, reducing the stressed area but, because of the reduction in the stress concentration factor, the reduced diameter bolt (as shown in Figure 12.8) is actually stronger in fatigue than the original.

Identification symbol (ASTM A 193)	Grade	AISI equivalent	Tensile strength (ksi, min)	Yield strength (ksi, min)	Size range (inches)
B5	5% Chromium	Type 501	100	80	Up to 4
B6, B6X	12% Chromium	Type 410	110 / 90	85 / 70	Up to 4 / Up to 4
B7, B7M	Chrome-moly	4140, 41, 42, 45 up to 4145H	125 / 100	105 / 80	Up to 2.5 / Up to 2.5
B16	Chromemoly-vanadium	None	125 / 110	105 / 95	Up to 2.5 / 2.5 to 4
B8	Chrome-nickel	Type 304 solution treated	70	30	All
B8	Chrome-nickel	Type 304 – sol'n treated and strain hardened	125 / 115 / 105	100 / 80 / 65	Up to ¾ / ¾ to 1 / 1.1–1.25
B8C	Chrome-nickel	Type 347 solution treated	70	30	All
B8M,	Chrome nickel-moly	Type 316 solution treated	80	35	All
B8M	Chromenickel-moly	Type 316 – sol'n treated and strain hardened	110 / 100 / 95	95 / 80 / 65	Up to ¾ / ¾ to 1 / 1.1–1.25

FIGURE 12.7 High temperature inch-series alloy and stainless fastener specifications.

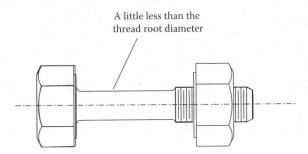

A little less than the
thread root diameter

FIGURE 12.8 A bolt designed for high-fatigue operation.

How much can the bolt body be reduced? Any reduction in the body diameter below the minor diameter of the threads reduces the static-load-carrying ability of the bolt. In most cases, this isn't as important as the dynamic (fatigue) load capacity, and Peterson's book covers these in detail. It is interesting to see that the bolt body diameter can usually be reduced significantly below the thread root diameter before the fatigue strength is reduced. However, there are several points that should be well recognized when considering reducing that diameter. They are:

1. When a bolt is turned down, careful attention has to be paid to the radius under the head, the radius at the threads, and the surface finish of the body. The new radius under the head should be at least as large as the original radius (It can actually be larger because of the increased diametrical clearance in the washer and hole.) and must be smooth and well blended. The radius leading from the body to the threads should be about the same as the head radius. Leaving tool or chatter marks and/or surface irregularities perpendicular

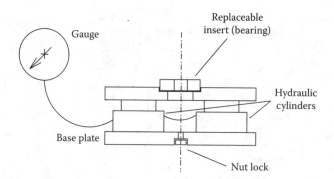

FIGURE 12.9 A device for analyzing the torque-tension relationship for multiple bolting conditions.

to the axis of the bolt will introduce additional stress concentrations. The machinist has to do a quality job.
2. Most bolts under 2.9 cm (1.125 in.) have rolled threads. These rolled threads are actually stronger than cut threads in fatigue applications because they have residual compressive stresses. Cutting into the bolt body can remove the compression forces but if it is machined as outlined in the preceding paragraph, there should be no net reduction in strength.

Assembly Practices

Using the patented device shown in Figure 12.9, we have conducted hundreds of analyses on a variety of fasteners, reviewing some of Dr. Bazergui's findings. One of these tests showed that lubricant on threads can reduce the torque needed to develop a given clamping force by an average 38%. Another series of tests showed that lubrication results in much more uniform tightening, i.e., there was far less scatter in the torque vs. tightness data for the lubricated bolts. A third series of tests showed that reusing bolts that have not been distorted or visibly damaged rarely has a significant effect on the clamping force. Based on these tests (and other data), we recommend lubricating all nuts and bolts, unless the equipment manufacturer specifically recommends otherwise.

This difference between lubricated and unlubricated bolts is especially obvious when stainless steel or aluminum bolts are used. Tests on the demonstrator show that an unlubricated stainless or aluminum bolt will typically only develop about half the clamping force of a lubricated one before the bolt seizes to the nut. After the bolt seizes (galls), it usually has to be cut apart because the two pieces have welded together and disassembly is impossible.

Bolting Patterns and Sequences

Almost every critical assembly, in which tightening of multiple bolts is involved, requires a procedure specifying the steps in which the bolts should be tightened and a sequence such as that shown in Figure 12.10. The goal of these bolting patterns is to reduce distortion of the bolted members and result in more uniform clamping force and more reliable assembly life. But, it is also an important point to be certain that the stresses on the joint are consistent with the design and don't include serious eccentric forces.

For good piping reliability, the flanges have to be properly aligned before the bolts are assembled and tightened, and this means you shouldn't use a come-along to pull the flanges into alignment and the flange faces should be parallel. Current industry practices (and ASME Boiler and Pressure Vessel Code) call for:

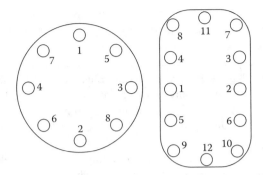

FIGURE 12.10 Two common tightening sequences.

- Flange flatness — Must be within 0.15 mm (0.006 in.). The maximum allowable out-of-flatness shall not occur in less than 20° of arc.
- Flange alignment — The two mating flanges must be parallel to within 1 mm in 200 mm (1/16th in./ft) and the centerlines of the two flanges must be within 3 mm (1/8 in.) maximum offset.

The value of these standards is that they improve long-term flange and piping system reliability.

There are many different tightening sequences, and for important assemblies, we generally recommend the following:

1. Inspect the fasteners to ensure all meet the same specifications.
2. Liberally lubricate the bolts, both on the threads and under the side that will be rotating.
3. Lightly snug the bolts to ensure good alignment and seating.
4. Tighten the bolts in a sequence to 60% of the maximum recommended torque, then 80%, and then 100%.
5. Repeat the sequence at 100% of the rated torque.
6. Go around the assembly twice, clockwise, tightening to the specified torque.

This bolting practice has been used in general chemical and petrochemical plants for over 50 years and, as far as we can determine, it was developed empirically and has produced very good results.

The only formal study we know of where clamping forces were accurately measured on a gasketed multiflange steel-piping assembly was conducted by the University of Dayton for the ASTM. In that outstanding experiment, a pair of 450 mm (18-in.) flanges and a wafer valve (with gaskets) were bolted together to try to understand the actual effect that tightening one bolt has on the clamping force of the others. The bolts were equipped with strain gauges that read out on a computer screen, and then tightened to prescribed clamping forces. As each bolt was tightened, the changes in the stress within the others could easily be seen. In the initial trial, the bolts were tightened in three steps, lightly snug, 80% of rated clamping force, and 100% of clamping force. When the 100% round was completed, they found that the ratio of clamping forces between the tightest and loosest bolts was 9:1. Following another round of tightening to 100%, still in the cross pattern, produced a ratio of 5.5:1. Then two successive circular paths resulted in a ratio of about 2.5:1. We suggest the six steps shown in the preceding text because, in industrial practice, it is typical that torque wrenches are used, not strain gauges, and the ability of torque to measure clamping force is much poorer than that of the strain gauges.

What is truly important to realize is that nonuniform tightening will bend flanges and result in leaks. The difficulty in recognizing this is that the leaks often don't show up until sometime later when the gasket can't cope with the errors, the deformation, and the residual stresses caused by the poor tightening practices. A leaking flange is a failure just as a broken bolt or a failed bearing, and the most common causes of leaky flanges involve improper tightening practices by people who don't understand how to do the job correctly.

A story that should go a long way toward illustrating the importance of careful tightening procedures happened many years ago at the plant where I worked. Early one morning, I was about 25 m (80 ft) off the ground on top of a vessel, measuring the flows into it as part of a reliability project. We had a well-intentioned but hard-of-learning mechanic who was given the assignment of replacing a 30-cm (12-in.) cast-iron pipe elbow. The mechanic walked up and began installing the cast-iron elbow in some steel piping. From far above and a good distance away, while I was measuring the flows, I watched him rig the elbow into position, put a bolt on one side at the 3 o'clock position, and then put another bolt at the 9 o'clock position. He then picked up an impact wrench and first tightened one of the bolts, then the other. As he was tightening the second bolt, the whole piping system jumped as the flange broke. It was horrendously noisy, there was a long way between us, and there was little I could do.

About half an hour later, the mechanic returned with another cast iron elbow. He rigged it into position, put a bolt in one side, and put a bolt in the other side. He then picked up his impact wrench and tightened first one bolt and then the other. Again, as he was tightening the second bolt, the piping system jumped as the flange broke off the elbow. I shook my head, silently wondered how some people even survived, and went back to measuring flows.

Another half hour later he came back with another 12-in. short-radius cast iron elbow. When I saw him returning, I decided that was enough flow data and started to climb down to talk with him.

We hear stories like this, and there are several common reactions:

1. We have all worked with people like these, and we chuckle or laugh about the things they've done.
2. We all know that when the cast iron flanges broke, there was probably enough stress to bend the mating steel flanges just a little.
3. We all know enough not to bolt up a cast iron flange like he did. But we all regularly see examples of people using impact wrenches to tighten flange bolts, one after the other, going around in a circle, and making no pretense of trying to use good bolting practices.

Until we think about it, few of us recognize that there will be problems and probable leaks the next time that flange is taken apart and rebolted. Without looking at the flange flatness, we might blame the people who just put the flange together for using poor practices.

Before leaving the subject of bolting practices, we should review why thread lubrication is an important point. Earlier in the chapter we said, "We recommend lubricating all nuts and bolts, unless the equipment manufacturer specifically recommends otherwise." Sometimes, as with automotive wheel studs, the manufacturers state that no lubricant should be used. In other cases, they may specify a certain lubricant.

We strongly recommend following the manufacturer's recommendations but in other applications, those where the manufacturers don't state specifically how to lubricate a threaded fastener (and this probably includes 99% of the bolts used in the maintenance world), the specific lubricant is not anywhere near as important as being consistent in the bolting procedure. In the first few pages of this chapter, we mentioned that about 82% of the torque goes into overcoming friction while 18% goes into stretching the bolt. If that coefficient of friction is increased by 10% (for example, from 0.120 to 0.132), it means 90% of the torque would go into overcoming friction and the tightening force would be reduced by more than 40%. In the design of a bolted joint, such as a steam or process line flange, the factor of safety on the bolt strength is almost always a very, very conservative number. What invariably causes the problem isn't that the bolts are too weak, it is that they aren't tightened uniformly. Being consistent with a lubricant eliminates one of the possible errors.

There are almost an infinite number of combinations of base metals, plating, and lubricants and unless there are specific instructions to the contrary, for the best reliability, we recommend standardizing on an assembly lubricant and applying it liberally.

FASTENER FAILURES

Fasteners fail mechanically either from overload or from fatigue. With overloads, the bolt fails immediately as the load is applied. The actual fracture may be ductile, with lots of distortion, brittle with no visible distortion, or somewhere in between. Photo 12.2 shows the ductile elongation of a tie bolt from the stator of a large motor, and if it had been tightened a little more, the bolt would have broken in a classic cup-and-cone ductile elongation failure.

Overload failures usually happen as the bolt is installed, and most are the result of overtightening, but they can occur at any time if the bolt is subjected to severe loads. Water hammer, dropped parts, and thermal shock loads are all examples of situations in which an installed, well-designed, and successfully operating bolt can fail because of overload.

Depending on the material, the failure usually looks like the bolt above or a classic brittle fracture with chevron marks pointing to the origin. However, there are also failures where the fastener is sheared, i.e., literally cut in half. These overload failures can be recognized by their deformation in the direction of the shear force.

FATIGUE FAILURES

Fatigue is by far the greatest cause of fastener failures and happens when the clamping force developed by a bolt is less than the separating force applied to it. The concept that a bolt has to be tightened to the point where the developed clamping forces are more than the separating forces

PHOTO 12.2 Note the plastic deformation on this motor stud.

is sometimes difficult to understand, because one of the fears people have is that a little more load on the bolt will then result in fracture. However, although the spring constant of the clamped materials must be taken into effect, from a practical standpoint, the bolt only sees a measurable change in stress when the separating force exceeds the clamping force.

Two solid verifications of the value of tightening until the clamping force is greater than the separating force can be seen in the following well-documented projects. In one case, an ultrasonic flaw detector was used to routinely monitor a group of 28 large cyclically loaded bolts. Over the first year of the project, we found several cracks and developed the ability to predict their growth. Then we found another bolt, one with a very small crack. It didn't have to be changed right away, and maintenance was backlogged. We continued to run the machine for several months while the crack slowly grew larger. Eventually, we got a new hydraulic torque wrench and tightened the bolt to the required 17,500 ft.lb. After this retightening, the machine ran for an additional 30 million cycles, and several inspections found no additional crack growth before the bolt was changed.

The second example involves a pump shaft assembly in which the impeller was held on by a stud and nut. Over many trials, we found that tightening the nut to 25% of the rated torque produced an assembly that would consistently fail in about 30,000,000 cycles. Tightening to 50% resulted in a life of about 100,000,000 cycles, whereas tightening to the full design value resulted in an assembly that never failed.

These examples are essentially the same and effectively prove how important it is to develop a clamping force greater than the separating force if fatigue failures are to be avoided.

In inspecting bolt failures, one of the clues to inadequate clamping force is that there is movement between the bolts and the mating pieces. The two examples shown in Photo 12.3 and Photo 12.4 have both failed from fatigue, and the photos show areas where the evidence of movement is relatively obvious.

Photo 12.3 is of a body-fitted bolt. Looking at the shank, it can be seen that the half that is closer to the nut still has the original machining marks on it, whereas the next section has been polished smooth. This polishing is proof that there was movement between the bolt and the pieces that were supposed to be tightly clamped together. In many similar examples, especially with electrogalvanized bolts, the side of the bolt body will show an irregularly blackened and discolored fretted area.

In a similar manner, the piece shown in Photo 12.4 is the nut from a failed bolt. The fracture was the result of a fatigue crack starting at about the 1:30 o'clock position. The contact surface on the underside of the nut shows two polished areas, one at about 10:30 and one at 2 o'clock. The only way these surfaces could have become polished during operation was if the mating surface was irregular and there was movement between the two pieces.

One of the contributors to the failure in Photo 12.4 was the irregular contact surface, and the importance of a well-lubricated, smooth contact cannot be overstated. After a bolt is tightened, it will normally lose a portion of its clamping force over time because of minute smoothing of the asperities and plastic deformation. If the contact area is small, not only will the forces be extremely high but also there may be substantial elastic deformation.

PHOTO 12.3 Inspection of the shank of this body fitted bolt shows movement and wear on the right side.

PHOTO 12.4 The face of the nut shows uneven contact and wear at both the 10:30 and 2 o'clock positions.

PHOTO 12.5 A fatigue fracture on a bent high-strength bolt.

The Grade 8 bolt shown in Photo 12.5 is an extreme case of this but does a good job of illustrating the point. It was used on a hydraulic pump and fractured after being in operation for several months. The photo clearly shows a fatigue failure with several ratchet marks and prominent progression marks. Inspection of the bolt body shows it was bent by about 0.6 mm (0.025 in.), and there is visible fretting on the shank just below the origins. The bend concentrated the load on one side of the head and the fact was that, even though the bolt was tightened, the load was concentrated at one point. The bolt flexed during peak load spikes and fatigue cracking resulted. (This bolt has been electroless nickel plated to preserve the surface features.)

HYDROGEN EMBRITTLEMENT AND HYDROGEN-INFLUENCED CRACKING (HIC)

At least when it comes to materials, I'm a firm believer that there is no such thing as a free lunch. The too common plant solution for a bolt failure is to replace the failed bolts with stronger ones but there is always a trade-off for higher-strength bolts. They are much more affected by corrosion, and HIC is more common in high-strength materials.

The terminology used to describe hydrogen damage leads to confusion. The most common term used by fastener people is *hydrogen embrittlement* and, when they say it, they talk about failures because of the absorbed hydrogen during processing or use, which then causes cracking. In the eyes of corrosion engineers and scientists, this type of cracking is HIC. They reserve the term hydrogen embrittlement for those instances in which the hydrogen causes a change in the chemistry of the metal, for example, when hydrogen turns the metal into a metal hydride.

We don't care what words you use as long as you recognize that there are always trade-offs for higher strength. In those situations in which a higher-strength bolt is used, the higher-strength bolts usually cost more, almost always are more susceptible to corrosion and the resultant fractures, and usually have less impact-absorbing ability than the lower-strength bolts. There is general agreement among fastener experts that bolts harder than HRC 38 shouldn't be used in any corrosive applications because of their tendency toward HIC. In the chemical industry, there are numerous prohibitions against using bolts harder than HRC 25 in applications involving exposure to acids because of possible HIC. Grade 8 bolts can be harder than HRC 38. Grade 5 bolts can be harder than HRC 25. These are two examples in which the use of higher-strength bolts can lead directly to fastener failures.

From a metallurgical standpoint, the problem is that the absorbed hydrogen causes a stress concentration within the metal at a microscopic crack tip. With static stresses, there is a reduction in area but with fatigue stresses the ensuing stress multiplication frequently causes failures. The mechanisms of hydrogen embrittlement and HIC are discussed in Chapter 7, and their identification requires detailed metallurgical analysis.

FAILURE LOCATIONS AND BOLT DESIGNS

Figure 12.11 shows the failure locations and is the result of a study done about 50 years ago. The failures occur in these areas because of stress concentrations and the causes and the failure locations haven't changed much over the years.

> The concept of stress concentrations was discussed in Chapter 5 and earlier in this chapter. It is one that everyone involved with fasteners and fastener problems should thoroughly understand because they cause the increase of local stresses in a part. They are caused by changes in metallurgy, changes in shape, shrink fits, and material defects called *stress risers*. The consequence is that the actual stress is much greater than the average stress in the part in the immediate area of the stress riser. The technical symbol for the stress concentration factor is K_t, and the factors for various thread shapes usually fall in the range of 4 (for well-rounded shapes) to 8 (for acme and stub acme threads).

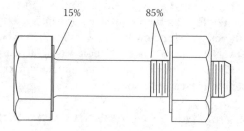

FIGURE 12.11 Most common locations for fatigue failures.

General Comments and Cautions on Bolting

- *Welding on fasteners:*
    - Unhardened bolts can be safely welded with no significant metallurgical changes.
    - Welding on heat-treated bolts, such as Grade 5 and higher, will change the metallurgy of the bolt. Is the new bolt stronger or weaker? Is there a stress concentration at the end of the rehardened zone? How serious are the residual stresses? Has hydrogen been introduced? These are some of the metallurgical questions that have to be answered if a welded heat-treated bolt is going to function reliably.
    - In addition, what effect has the heating had on the bolt preload? Has it reduced the preload so that an assembly that started out with adequate clamping force now sees significant fatigue stresses? In tests on the bolt demonstrator, several times a bolt was tightened to the rated preload, then lightly tack welded on the head. The heat and resultant elongation resulted in permanent preload reduction of about 50%.
- *Reuse of nuts and bolts:*
    - Some people and some books advise not to reuse a nut and bolt, but a reused bolt can be as safe as a new one if you check to see that there is no visible rust or corrosion and the threads are not distorted. How should you check for thread distortion? Run a nut down the threads with your fingers and it shouldn't bind. (Sometimes the cost of inspection is more than the bolt cost. However, if a bolt isn't distorted or corroded, our opinion is that it can be reused.)

FAILURE EXAMPLES

Example 12.E1

Photo 12.E1 shows a U.S. thread, 3.7-cm (1.5-in.), Grade 8 bolt used to hold a large platen on a press and it operated in a high-humidity atmosphere at about 25°C (77°F). There are several of these bolts that hold the platen in position and they can be impact loaded if a large piece is dropped onto the platen. The plant has had a history of these failures over the past few years, and they are in a location where routine inspection is extremely difficult.

PHOTO 12.E1 Reference Example 12.E1.

Inspection

A close examination of the fracture origin, centered at about 7 o'clock, shows some small corroded and discolored pits. The fracture surface is uniform in smoothness up to the shear lip that runs from about 3 o'clock to 9 o'clock. Shear lips are common on tension failures and occur when the crack changes direction as it grows across the piece. This change in direction happens because the load center is no longer in the center of the piece. On shadowing the fracture surface it can be seen that there are a series of chevron marks that point toward a relatively broad origin. Examination of the contact surface of the head showed no polishing, fretting, or other evidence of movement. This is a low-alloy steel bolt, and there is rust on it.

Diagnosis and Recommendation

This is a brittle overload failure and happened as the load was applied. We recommended that a metallurgical examination of the bolt be conducted before any other action. This would determine if the fracture was influenced by hydrogen damage. Whether it was or not, look at the operating practices to understand the loading and why it occurs. If corrosion influenced the failure, a specially coated bolt should be considered.

EXAMPLE 12.E2

Photo 12.E2 is a nut from a 13 mm (½-in.) B8 stainless steel bolt. This is part of an SS bolt assembly that is used at ambient temperatures in a damp atmosphere, and it failed suddenly during assembly.

Inspection

The fracture face is almost flat and perpendicular to the major axis. The face shows the swirl marks and eccentric eye that are typically seen in a ductile overload torsional failure. In addition, the threads in the nut are deformed, and there was no evidence of lubrication.

Diagnosis and Recommendation

It appears to be a ductile overload failure that occurred during tightening when the threads seized. If there is a question about whether it occurred during assembly or removal, a metallurgical

PHOTO 12.E2 Reference Example 12.E2.

examination can be made. That will show a deformed layer along the surface, and the direction of the deformation will answer the question. We recommended that future bolts be lubricated with a good grade of extreme pressure (EP) lubricant.

EXAMPLE 12.E3

This is one of a group of 3.5 cm diameter (1½-in.) fine thread bolts used to hold the upper frame to the base of a massive press in a forging shop. The material meets the specifications for a B7 bolt, and the bolt in this location had broken and had been replaced in the past. The plant is located in northeastern U.S., and the operating atmosphere varied from generally being close to outdoor ambient on weekends to being 10 to 20°C (18 to 36°F) warmer during the two shifts of the operating week.

Inspection

The fracture face shows a fatigue failure starting at about the 8:30 o'clock position in Photo 12.E3A. In the very beginning, there are several progression marks, then there is another progression mark about a quarter of the way across the face, with the instantaneous zone (IZ) occurring at the point where ¼ of the surface remained. The entire fracture face is essentially perpendicular to the long axis of the bolt and, as shown in Photo 12.E3B, the tips of the threads have been worn smooth. There is a great deal of grit and carbon dust on the entire bolt.

Diagnosis and Recommendation

The progression marks and change in surface roughness show that the bolt failed from fatigue. There are several ratchet marks but the change in surface color shows they occurred after the cracking had started. The ratchet marks indicate a high stress concentration, and they are common with a threaded fastener. The rounded tips of the threads show the base plate has been rubbing against the bolt, and the press pieces were moving side to side relative to each other. Based on these, our opinion is that the failure was from low-cycle fatigue caused by occasional peak stresses from very heavy forging loads. The recommendation was that they remove, inspect, lubricate, and retighten the remaining bolts using several steps in the procedure. In addition, if additional failures occur, it may be that the mating surfaces are no longer flat to each other, and careful shimming is needed.

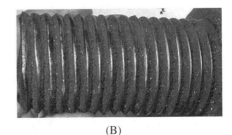

(A)   (B)

PHOTO 12.E3 Reference photos for Example 12.E3 (A and B).

PHOTO 12.E4 Reference photo for Example 12.E4.

EXAMPLE 12.E4

Photo 12.E4 shows cracking in the underhead radius of one of a group of 6.1-cm (2.5-in.) bolts that anchor the lower die set of a 500-t forging press to the machine frame. The machine runs in a heated and controlled atmosphere and does not see the variations of the previous example. There are eight bolts that hold the die set in position, and we had been asked to nondestructively test several sets of bolts. They had been using a hydraulic torque wrench to tighten the bolts.

Inspection

A magnetic particle inspection found cracks in about 35% of the inspected bolts. We then looked carefully at the machining. It appeared that the bolts had been machined by different shops from different materials and that there were different radii under the bolt heads. Some of the underhead radii were very generous, whereas those on the cracked bolts were small and machined in obvious steps. In addition, hardness testing of the bolt materials showed that the uncracked ones were about HRC 30, whereas the cracked ones were HRC 20. After we looked at and found the cracks, one of the maintenance personnel found a broken bolt that had been left in a corner and rusted, but that fracture was obviously from fatigue.

Diagnosis and Recommendations

The bolts failed from fatigue, indicating that either they had not been tightened correctly or the machine frame was flexing. However, the replacement bolts they were using had been made from a sample rather than a drawing, and:

1. The underhead stress concentration was higher than it should have been.
2. The material had only 66% of the fatigue strength that it should have had (HRC 30 vs. HRC 20).

We recommended that they change the material specification back to the original and ensure that any replacement bolts are machined to the same specifications as the originals. We also recommended that they go around the die set bolts in sequence at least twice in the tightening procedure.

EXAMPLE 12.E5

This is a bolt from a hydroelectric plant runner coupling on a shaft that turned at about 300 r/min. (The runner in a hydroelectric turbine is the propeller-like device that is turned by the water, and

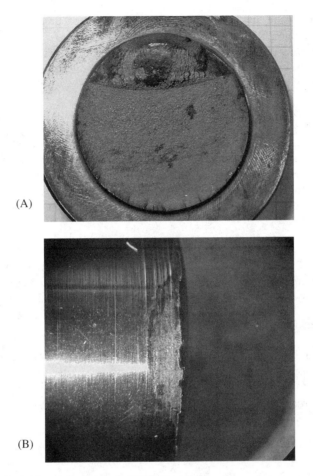

(A)

(B)

PHOTO 12.E5 Reference photos for Example 12.E5 (A and B).

the coupling is a rigid coupling between two shafts.) This 4-cm (1.5-in.) body-fitted bolt assembly met the requirements of ASTM A 193 B7 and had been in essentially continuous operation for 8 years when it fractured.

Inspection

The fracture face shows classical fatigue failure characteristics. The failure started at the very bottom of Photo 12.E5A where there are several ratchet marks. The fracture face was flat, and the crack gradually grew across the body, getting progressively rougher and rougher until the final fracture occurred about 70% across the shank. Looking at the underside of the bolt head found polishing (from the 10 o'clock to the 11:30 o'clock position in the photo) where there had been movement between the bolt and the coupling face and the machining marks were worn off. Then, looking at the side of the bolt in Photo 12.E5B, some fretting can be seen immediately under the bolt head.

Conclusions and Recommendations

Based on the observations, the bolt had been loose in the coupling assembly for some time, the head contact was not good, and a fatigue crack started where it had been fretted. When the assembly is to be reassembled, our recommendations were to:

1. Check to make sure the fit of the coupling assembly is secure.
2. Check the bolt and bolt hole perpendicularity.
3. Lubricate the bolt threads and the contact face under the nut with an antiseize compound and tighten the two coupling bolts to 1250 ft.lb in four steps, then recheck the final torque reading.

EXAMPLE 12.E6

Photo 12.E6 is an igniter from a large gas engine that is similar to a spark plug used in a typical automotive engine. The site had many of these engines and, when an igniter failed, they had to be replaced. However, frequently during attempts at removing the old igniter, the threads would seize and gall, requiring lengthy and expensive repairs. The problem had not been seen with the OEM igniters but these were a less expensive replacement and not quite the same construction.

Inspection

The inspection found the threads were essentially being welded into position. This galling is not uncommon on assemblies that see severe temperature cycling. Hardness testing of the steel, male and female thread assemblies found they were essentially the same hardness.

Conclusions and Recommendations

The igniters were galling because of the compatibility of the materials when subjected to heat and pressure. (The two pieces were welded together because the materials were so similar.) We suggested they could go back to the OEM igniters or they could speak with the new supplier and ask that they change to a harder threaded body and consider coating the threads with a material that did not weld to steel.

EXAMPLE 12.E7

Photos 12.E7 A and B show two bolts that were used to support a hoist in a manufacturing area. The facility was used for light manufacturing in the New York–Baltimore industrial complex, and the room where the hoist was used was generally between a low of 10°C (50°F) and a high of 35°C (95°F). The hoist was used near a hydrochloric acid system and there were occasional traces of acid fumes in the air. It was cycled about 100 times per month and the weight of the lifted load was less than 5% of the bolt's static capacity.

PHOTO 12.E6 Reference Example 12.E6.

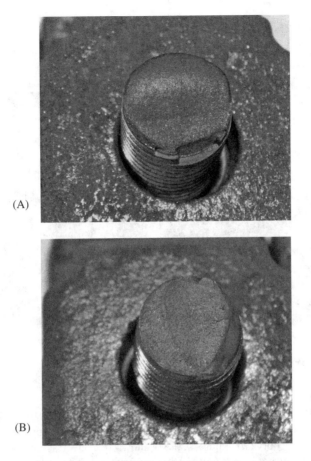

(A)

(B)

PHOTO 12.E7 Reference photos for Example 12.E7 (A and B).

Inspection

The two hanger bolts were aligned above the hoist as shown in the two photos with the hoist support midway between them. Examining the first bolt shows that it failed from fatigue with the crack starting at the top of the photo and growing downward, across the hoist axis. In addition, there was a substantial amount of rust on the mounting block, the piece shown in the background. The second bolt showed a fatigue crack that worked its way about a third of the way across the bolt before the catastrophic failure occurred. This bolt was much more heavily stressed than the first one, and the lack of corrosion on the mounting block shows this fracture happened immediately before the hoist fell.

Conclusions and Recommendations

From the direction of the crack growth on the first bolt, it is apparent that the crane was side loaded. The first bolt failed from these excessive loads, and then the second bolt was carrying the full load and subjected to a bending moment. It eventually failed from fatigue.

EXAMPLE 12.E8

This is a 5-cm (2-in.) Class 10.9 bolt from the counterweight of a piece of mining equipment. There are several of these bolts that hold the weight in position, and the outboard ones periodically fail. Loss of the counterweight could be very dangerous because the machine could tip instead of lifting.

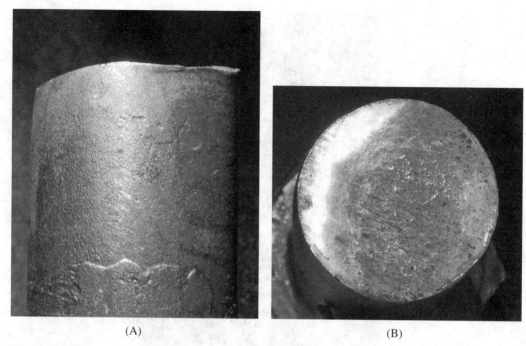

(A)   (B)

PHOTO 12.E8 Reference photos for Example 12.E8 (A and B).

Inspection

As can be seen in Photo 12.E8A, a straightedge placed along the side of the bolt shows that it is bent off to one side. Then, looking at the opposite side, it can be seen that the edge of the fracture face has been deformed outward. Photo 12.E8B shows that the fracture surface is very flat and perpendicular to the body, but it isn't uniform in smoothness and doesn't have any signs of a fatigue failure's IZ. A close examination shows the surface features are all distorted in the same direction as the overhang shown in photo A.

The bolts showed no evidence of looseness.

Conclusions and Recommendations

The fact that there is deformation shows this is either a very-low-cycle (VLC) fatigue or an overload failure. There is no distortion in the direction of the long axis of the bolt. Therefore, it wasn't an axial load that caused the failure, it was a shear load. The smoothness of the fracture surface is indicative of a bolt that experienced a tremendous side force, essentially cutting the bolt in two. This was not a fatigue failure so the forces that caused it were applied immediately before the failure occurred.

We suspect that these counterweight bolts were being impact loaded by unusual strains on the machine but did not have other adequate support. We did not get to see the specific machine in which the failures were occurring but did watch similar machines and they had the potential for occasional shock loads. We recommended they:

- Talk with the factory design personnel to see what the FEA shows happened during heavy loading, i.e., what the magnitude of the actual loading was.
- Keep a log of failures including who was operating the machine and what the machine was being asked to do.
- Add the bolts to the preshift inspection procedure.

13 A Brief Review of Miscellaneous Machine Element Failures

CHAINS

In understanding chain failures, we should probably start by pointing out that most roller chains are built to ASME standards that were written many years ago. These standards describe the physical properties of the chains but do little to describe their quality. There are some manufacturers who build chains as a commodity item to meet the minimum specifications, and there are others whose products offer much more than meeting a minimum standard.

CHAIN DESIGN

Most chains used in industry are roller chains. Their basic construction is almost the same as the chain of a bicycle, and the components are shown in Figure 13.1. All of the contact components are hardened, with most of them being case hardened. In addition, there are "engineered chains" that share many of the same design criteria.

Chain drives designed "by the book" are typically designed to last from 15,000 to 20,000 operating hours before the chain is replaced. However, lubrication is critical to long chain life because of all the relative movement and contact. Other important influences are:

- *Sprocket sizes* — The more the number of teeth on the sprocket, the less the chain flexes and wears.
- *Slack* — Excessive slack will result in more chain movement and wear.
- *Misalignment* — This can concentrate the forces on one side of the chain instead of having the two paths work together.
- *Sprocket wear* — As the sprocket wears, the pitch changes and slowly reduces the chain life.
- *Corrosion* — The chain components are hardened and, with the effect of HIC on hardened steels, corrosion can cause chain breakage.

LUBRICATION

The chain components that rub against each other are case hardened to above HRC 50. As the chain flexes during operation, these pieces wear, and eventually the wear particles form a lapping compound. Lubrication provides two functions:

1. With lightly lubricated components, when lubricants are applied by an occasional brush or drip application, the lubricant helps to reduce the wear.
2. With spray or dip application, the lubricant actually tends to wash out the wear particles.

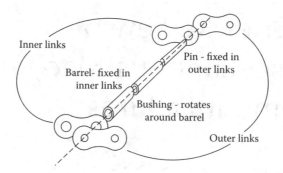

FIGURE 13.1 The components of a roller chain.

A study by Rexnord, Inc., showed the following effects of lubrication:

Chain Operating Conditions	Relative Chain Life
Corrosive and/or abrasive	Less than 1
Clean and dry	1
Manual (intermittent) lubrication	10
Oil bath	100–500
Oil stream	1,000–10,000

The lubricant used should be a good grade of EP oil, and if the chain is being used in a wet atmosphere, it should also have additives for improved corrosion resistance. A general guideline for lubricant viscosity is that it should be between 60 and 120 cSt at the operating temperature.

WEAR

Regardless of how much the chain is lubricated, it is eventually going to wear and a gauge can be used to check the relative condition of the chain. In the active part of the chain, set up a measure over some convenient distance, such as two pins that are 100 cm (25 in.) apart, then periodically remeasure the distance between the pins. When the elongation exceeds 3%, the wear rate will increase dramatically because it has worn through the hardened case.

The chain wears every time it flexes, and a chain with a lot of slack and frequent turns, as shown in Photo 13.1, will wear out much more rapidly than one with only a slight sag.

Sprocket wear is important because of the effect on chain stress. As the sprocket wears, the pitch changes and, if a new chain is run on worn sprockets, this will concentrate the entire load on the last link or two as opposed to distributing it over several links. The increased load will increase the wear rate and reduce chain life. One way to try to counteract this is to harden the sprockets, and our plant activities indicate that in most applications, hardening the sprockets to about HRC 45 or so will improve system life when compared to the industry standard HRC 35 sprockets.

PHOTO 13.1 A chain wears every time the links flex. There are 10 wear points on this chain compared with four on a properly adjusted one.

How much sprocket wear is tolerable? That really depends on the nature of installation (the chain size, the sprocket size, and the application) and the cost of the chain. On small chains (bicycle sized) or small sprockets, almost any visible wear will cause skipping of the chain. In most industrial roller chain applications, 1/8 in. wear is excessive.

> The chain and sprocket wear on our bicycles gives an interesting insight into this question. We have a tandem bicycle and do a lot of touring with it. Because it uses standard chain and sprocket components, the drive to the rear wheel wears fairly quickly, much, much quicker than on either of our individual bikes. Changing the rear cluster is about four times as expensive as changing the chain, and it is very time consuming. As a result, we've decided to change the chain when the wear is about 1% and have appreciated an increase in sprocket life. One other point that can be seen from bicycle applications is an appreciation for minimum number of tooth specifications. As the rear sprockets become smaller, the wear rate increases dramatically.

CHAIN BREAKAGE

We have never seen a chain break when there haven't been excessive loads or obvious corrosion.

Photo 13.2 is of a link from a size 160 chain, and looking at it, several interesting points are immediately apparent:

1. The end of the pin shows that it is case hardened with an thin case around the entire exterior.
2. The fracture face on the end of the pin shows a fatigue failure that starts at about the 12:30 o'clock position and also shows that the pin was heavily loaded in bending.
3. The irregular wear on the side of this outer link shows there was substantial misalignment.

Some guidelines on chain breakage are:

- Pin failure:
    1. Look at the direction of the crack growth. Generally caused by worn or badly aligned sprockets.
    2. Could be an impact load that starts a fatigue crack that takes a long time to grow across the pin. When this happens, the IZ is usually very small.

PHOTO 13.2 The fatigue fracture of a case-hardened pin.

PHOTO 13.3 A born worn unhardened sprocket.

- Roller failure — Most likely, a result of a badly worn sprocket in which the teeth hit the rollers but can occur with a badly worn chain on a high-speed drive. The sprocket in Photo 13.3 was off a dirt bike. The owner put a new chain on about 5 h before the sprocket was removed, and then the chain was removed. In that time, a third of the rollers had been broken off the chain.
- Bushing failure — Usually the result of prolonged system overloads.
- Side link failures — Look at the failure face:
  1. Instantaneous fracture is almost always a single overload.
  2. Fatigue failure is usually the result of prolonged system overloads.
  3. HIC is from corrosion acting on the hardened steel.
- Cotter pin failures — These are almost always the result of misalignment. (Cotters are designed to hold the pieces in position. They are not designed to withstand thrust loads.)

The failure analysis procedure we use for a chain drive is:

- Layout and reassemble the failed pieces.
- Identify the failed components.
- Inspect the links to see if they were bent or deformed.
- If pieces were broken, was it fatigue, overload, or hydrogen-influenced cracking (HIC)?
- What is the wear pattern on the pins, rollers, and side plates?
- How well were the sprockets aligned?
- What were the actual lubrication practices?

LIP SEALS

The life of the current design of lip seals is generally in the range of 8,000 to 10,000 h. Beyond this, they can still provide a measure of sealing but can't prevent leakage. Two other approaches that offer much better life are:

1. The dynamic sealing devices offered by several manufacturers. They are much more expensive than a lip seal and usually require special machining of the housing but they last and protect the machinery essentially indefinitely.
2. The conventional lip seal manufacturers are starting to provide two-piece seals with an estimated life in excess of 20,000 h.

The conventional lip seal design is such that the lip rides against the shaft and the shaft surface finish is critical. Figure 13.2 shows the sketch of a shaft and the running surface specification from several major seal manufacturers. Of interest is that for optimum life, they recommend the shaft be hardened to at least HRC 30 because the seal lip tends to wear the contact area.

In any installation in which there is rubbing contact, lubrication is critical and:

• A well-lubricated seal lip will outlast a dry one by thousands of hours.
• At the typical surface speeds seen in industrial machinery, multiple lip seals are usually shorter lived than single lip seals because of the heat generated at the unlubricated second lip.

This second category involves Arrhenius' rule and the exponential increase in degradation rates as temperatures increase.

A challenge faced in all lip seal applications is that of the lip being flexible enough to meet the imperfect operating conditions. The conundrum is that the application can't have high runout, high speed, and high STBM (shaft-to-bore misalignment) tolerances (see Figure 13.3). For equal life, high misalignment means the application has to accept either lower speed or less runout. On the other hand, if it is a high-speed application, it can't have both high runout and high STBM, etc.

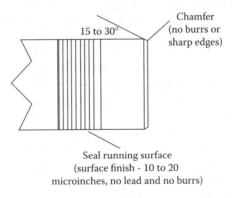

FIGURE 13.2 The shaft surface specifications for a lip seal.

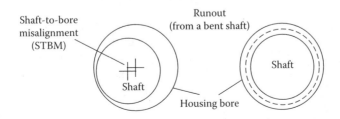

FIGURE 13.3 Comparing misalignment with runout.

PHOTO 13.4 The seal lip was cut during installation.

FAILURE ANALYSIS

As with other equipment, seal failures rarely result from just one cause. (There may be a single physical cause but there are always multiple human errors that lead to the failures.)

The procedure we usually use is:

1. Look at the conditions around the seal, including the operating environment and equipment temperature ranges.
2. Get the operating history, including how long it has been installed, who installed it, any machine data about load ranges and swings, etc.
3. Talk to the operators and maintenance people asking for their opinions on the source of the problem.
4. Using low-power magnification, look at the seal components, the lip, the housing, and the shaft surface.
5. From the preceding items, diagnose what happened.

The seal problems we've seen usually begin at installation when the lip is damaged or the seal shape is distorted. Photo 13.4 shows the lip of a seal used on a piece of medical equipment. The manufacturer was plagued by user complaints about oil and grease leakage from the equipment and was convinced that the seal manufacturer had a problem with their materials. It wasn't until they examined the seal lip using a 10× magnifying glass that they saw the small nick in the seal lip just at the horizontal centerline of the photo.

FLEXIBLE COUPLINGS

Gear and grid couplings are designed to accept a certain amount of misalignment as long as they are properly lubricated. At very low speeds, below about 200 r/min, lubrication can become complicated but above that the use of the specialty coupling greases should limit the need to relubricate a properly aligned coupling to the following intervals:

- Grid couplings — Every 5 years
- Gear couplings — Every 3 years

One of the interesting things about inspecting failed couplings is that the failure causes are relatively obvious. For example, looking at Photo 13.5, one can see:

PHOTO 13.5 An improperly lubricated and badly aligned grid coupling.

1. From the wear, i.e., the grid imprints on the inside of the cover, it is apparent that the hubs were seriously misaligned.
2. From the wear on the ends of the male grid driving pieces, the rounded surfaces on the right hub, it can be seen that the grid was tight up against them and that the center gap spacing was incorrect.
3. There was rubbing against the hubs on both sides, and the two paths from the seal retainer can be seen, but the seals were not installed.
4. Obviously it had been a long time since it was lubricated.

From this we know that the person who installed it didn't properly space it on installation and also caused a misalignment. The seals were also not properly installed. (From this flagrant series of errors and the lack of knowledge of the importance of alignment procedures, we would also suspect that the management is inadequate.)

In contrast to the preceding disaster, look at the coupling grid in Photo 13.6. This was in operation on a skip hoist for about 9 months and is in excellent condition but still tells an interesting story as follows:

1. The wear patterns are a little heavier on one side than the other, suggesting that the drive forces may not be quite balanced in both directions.
2. For 9 months of operation, the amount of wear is very small, and the coupling is well lubricated.
3. The polishing on the top of the grid shows that it is rubbing against the cover and that the coupling is misaligned.

The gear coupling hub shown in Photo 13.7 is badly worn. Inspection of the teeth around the hub shows that they do not appear to be uniformly attacked, indicating that the alignment is not good, and less than half of the tooth thickness remains. The coupling has been in use for about 2 years and was lubricated with "a good grade of grease." The combination of misalignment and improper lubricant has resulted in the premature death of another coupling.

Photo 13.8 shows 9 months of operation on the hub from a 530 kw (400-hp), 3600-r/min motor that drove a compressor until the motor failed. The hub shows substantial wear on the teeth and, similar to the immediately preceding example, the wear is not uniform around the hub, indicating there was serious angular misalignment along with the 0.4-mm (0.016-in.) parallel misalignment. This was in a small waste treatment plant, and the crew that installed the assembly didn't understand the alignment requirements and didn't have any grease — so they looked around until they "found something" to put in it. The compressor failed about a month later!

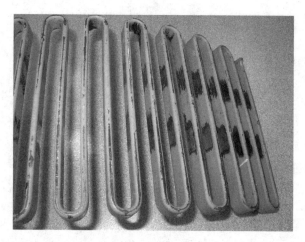

PHOTO 13.6 This coupling grid had been in operation for 9 months.

PHOTO 13.7 This gear coupling ran for 2 years while misaligned and improperly lubricated.

PHOTO 13.8 The misalignment of this coupling contributed to costly failures.

Bibliography for Practical Plant Failure Analysis

Neville W. Sachs

CHAPTER 1 — AN INTRODUCTION TO FAILURE ANALYSIS

Latino, R.J. and Latino, K.C., *Root Cause Analysis, Improving Performance for Bottom Line Results,* CRC Press, Boca Raton, FL, 1999.

Strive for Excellence ... the Reliability Approach, Reliability Center, Hopewell, VA, 1980 .

Kletz, T., *Learning from Accidents,* 3rd ed., Elsevier, 2001.

Kletz, T., *Lessons from Disaster,* 1993, Golf Publishing, Houston, TX.

Swain, A.D. and H.E. Guttman, *Handbook on Human Error Reliability Analysis,* NUREG/CR-1278, SAND80-0200, Sandia National Laboratories, August 1983.

CHAPTER 2 — SOME GENERAL CONSIDERATIONS ON FAILURE ANALYSIS

Wulpi, D.J., *Understanding How Components Fail,* ASM International, Metals Park, OH, 1985.

ASM International, *Case Histories in Failure Analysis,* ASM International, Metals Park, OH, 1979.

CHAPTER 3 — MATERIALS AND THE SOURCES OF STRESSES

Osgood, C.C., *Fatigue Design,* John Wiley & Sons, New York, 1970.

Metal Fatigue, Theory and Design, Angel F. Madayag, Ed., John Wiley & Sons, New York, 1969.

Three Keys to Satisfaction, Climax Molybdenum Company, New York, 1961.

Hertzberg, R.W., *Deformation and Fracture Mechanics of Engineering Materials,* 2nd ed., John Wiley & Sons, New York, 1983.

Design Guidelines for the Selection and Use of Stainless Steels, Nickel Development Institute, Toronto, Ontario, Canada, 1976.

CHAPTER 4 — OVERLOAD FAILURES

ASM International, *Metals Handbook: Properties and Selection*, Vol. 1, ASM International, Metals Park, OH, 1961.

ASM International, *Metals Handbook: Failure Analysis and Prevention,* 8th ed., Vol. 10, ASM International, Metals Park, OH, 1975.

Wulpi, D.J., *Understanding How Components Fail,* ASM International, Metals Park, OH, 1985.

Barer, R.D. and Peters, B.F., *Why Metals Fail,* Gordon and Breach, New York, 1970.

CHAPTER 5 — FATIGUE FAILURES: THE BASICS

Wulpi, D.J., *Understanding How Components Fail,* ASM International, Metals Park, OH, 1985.

ASM International, *Metals Handbook: Failure Analysis and Prevention*, 9th ed., Vol. 11, ASM International, Metals Park, OH, 1979.

ASM International, *Metals Handbook: Failure Analysis and Prevention,* 8th ed., Vol. 10, ASM International, Metals Park, OH, 1979.

Brooks, C.R. and Choudry, A., *Metallurgical Failure Analysis,* McGraw-Hill, New York, 1993.

Bannantine, J.A., Comer, J.J., and Handrock, J.L., *Fundamentals of Metal Fatigue Analysis,* Prentice-Hall, Englewood Cliffs, NJ, 1990.

Peterson's Stress Concentration Factors, 2nd ed., Walter Pilkey, Ed., John Wiley & Sons, New York, 1997.

CHAPTER 6 — FATIGUE FAILURES: FAILURE INFLUENCES AND FATIGUE INTERPRETATIONS

Wulpi, D.J., *Understanding How Components Fail,* ASM International, Metals Park, OH, 1985.

Barer, R.D. and Peters, B.F., *Why Metals Fail,* Gordon and Breach, New York, 1970.

ASM International, *Metals Handbook: Failure Analysis and Prevention,* 9th ed., Vol. 11, ASM International, Metals Park, OH, 1979.

ASM International, *Metals Handbook: Failure Analysis and Prevention,* 8th ed., Vol. 10, ASM International, Metals Park, OH, 1979.

CHAPTER 7 — UNDERSTANDING AND RECOGNIZING CORROSION

Uhlig's Corrosion Handbook, Winston Revie, R., Ed., sponsored by the Electrochemical Society, John Wiley & Sons, New York, 2000.

Chawla, S.L. and Gupta, R.K., *Materials Selection for Corrosion Control,* ASM International, Metals Park, OH, 1993.

Fontana, M.G. and Greene, N.D., *Corrosion Engineering,* McGraw-Hill, New York, 1978.

ASM International, *Handbook of Corrosion Data,* Bruce Craig, Ed., ASM International, Metals Park, OH, 1989.

ASM International *ASM Handbook: Corrosion,* Vol. 13, ASM International, Metals Park, OH, 1987.

ASM International *ASM Handbook: Corrosion*, Vol. 13A, ASM International, Metals Park, OH, 2003.

CHAPTER 8 — LUBRICATION AND WEAR

Association of Iron and Steel Engineers, *The Lubrication Engineers Manual,* 2nd ed., AISE, Pittsburgh, PA, 1996.

Wulpi, D.J., *Understanding How Components Fail,* ASM International, Metals Park, OH, 1985.

Klamann, D., *Lubricants and Related Products,* Verlag Chemie, Deerfield Beach, FL, 1984.

Spotts, M.F., *Design of Machine Elements,* Prentice-Hall, Englewood Cliffs, NJ, 1961.

CHAPTER 9 — BELT DRIVES

Belt Preventive Maintenance Manual, The Gates Rubber Company, Denver, CO, 1989.

Installation and Maintenance of V-belt Drives, Goodyear Power Transmission Products, Lincoln, NE.

D.E. Roos of Gates Rubber Company, Engineering with Belts, National Conference on Power Transmission, 1975.

Basics of Power Transmission Belts, *Plant Engineering Magazine*, Joseph L. Foszcz, Ed., McGraw-Hill, New York, October 1991.

Rubber Manufacturers Association, Various RMA/ANSI Standards, Rubber Manufacturers Association, Washington, D.C.

J.D. Shepard of Gates Rubber Company, Fundamentals of Belt Power Transmission, National Conference on Power Transmission, 1974.

Bloch, H.P. and Geitner, F.K., *Machinery Failure Analysis and Troubleshooting,* 3rd ed., Vol. 2, Gulf Publishing Company, Houston, TX, 1997.

Service Manual for Industrial V-Belt Drives, Carlisle Power Transmission Products, Miamisburg, OH, 2003.

CHAPTER 10 — ROLLING ELEMENT BEARINGS

NSK Americas Inc., NSK Technical Report Cat. No. E728d, NSK Americas, Ann Arbor, MI 1999.

NSK Americas Inc., Rolling Bearings, Cat. No. E1101c, NSK Americas, Ann Arbor, MI, 1999.

Society of Tribologists and Lubrication Engineers, *STLE Life Factors for Rolling Element Bearings,* Erwin Zaretsky, Ed., STLE, Park Ridge, IL, 1999.

Harris, T.A., *Rolling Bearing Analysis,* 3rd ed., John Wiley & Sons, New York, 1991.

Rolling Element Bearings and Their Contribution to the Progress of Technology, FAG Kugelfischer Georg Schafer KGaA, Schweinfurt, West Germany, 1986.

CHAPTER 11 — GEARS

Association of Iron and Steel Engineers, *The Lubrication Engineers Manual,* 2nd ed., AISE, Pittsburgh, PA, 1996.

AGMA Standards (various), American Gear Manufacturers Association, Washington, D.C.

Robert, L.E. and Jane, M., Hot to Analyze Gear Failures, *Power Transmission Design Magazine*, Penton Publishing, Cleveland, OH, March 1994.

Borden, D.L. and Holloway, G.A., Failure Analysis Gears — Shafts — Bearings — Seals, The Falk Corporation, Milwaukee, WI, 1978.

Timmerman, D.N., How to Specify and Select Gear Drives, *Coal Mining Magazine*, January 1985, Falk Reprint #10,561.

Gear Lubrication I and II, *Lubrication Magazine*, Texaco Inc., White Plains, New York, 1980.

CHAPTER 12 — FASTENERS AND BOLTED JOINT LEAKS

Handbook of Bolts and Bolted Joints, John Bickford and Sayed Nassar, Eds., Marcel Dekker, New York, 1998.

CHAPTER 13 — MISCELLANEOUS MACHINE ELEMENTS

Chain Installation and Maintenance, U.S. Tsubaki, Wheeling, IL, 1992.

W.F. Hofmeister of Rexnord, Inc., Lubrication of Drive and Conveyor Chains, National Conference on Power Transmission, 1982.

Maintenance Suggestions, Diamond Chain Company, Indianapolis, IN, 1981.

Index

G

Galvanic corrosion, 106–109
  appearance, 107
  area effect, 107
  galvanic series, 106
  prevention, 107
Gears, 197–223
  alignment monitoring, 206
  analyzing failures
      examples, 218–223
      procedure, 215–218
  backlash, 206
  case crushing, 215
  corrosion 210–211
  design life, 207
  designs, 198–200
  deterioration mechanisms
      surface hardened teeth, 204–205, 212–215
      through hardened teeth, 204–205, 208–211
  materials of construction, 202–204
  micropitting, 212
  operation, 197, 201–202
  pitting, 204–205, 208–211
  rippling, 213
  spalling, 215
  terminology, 197–198
  tooth
      action, 200–201
      contact patterns, 202, 204–206
      stress fluctuations, 201–202
  types, 198–199
Gear tooth micrometer, 210
Goodman diagram, 84

H

Hardness testing
  chain sprockets, 248
  rolling element bearings, 189
  suggested equipment, 24
Heat treating, 39–40
  problems, 215, 218
Hertz, Heinrich, 164
Hertzian
  contact region, 124–125, 164
      lubricant viscosity, 132
      viscosity conversion, 124–125
  fatigue
      gear tooth action, 200–201, 208
      rolling element bearing action, 181–183, 185
High carbon steel
  alloying elements, 35–37
  metallurgy 32–34
High-cycle fatigue
  chart, 58
  definition, 57
  diagnosing a failure, 60

interpretation
    charts, 73–75
    example, 72,76
  metallurgical effects, 60
  multiple origins, 67–70
  progression marks, 61–63, 65–67
  range of stress cycles, 58
High temperature direct conversion, 95
Hot rolled steel, 38
Human error, 2–7, 12, 13
  categories, 6–7
  distribution, 7
  finding, 22
  frequency, 3,7
  significant error (def.) 4
  significant error (study) 5–7
Human roots, 2, 4–8
Hydrogen damage mechanisms, 119–122
  hydrogen blistering, 119
  hydrogen embrittlement, 122
  hydrogen influenced cracking, 120–121
      appearance, 120–121
      effect of steel tensile strength, 120
      prevention, 121
Hydrogen embrittlement, 96
Hydrogen generation, *see* Atomic hydrogen generation
Hydrogen induced cracking, *see* hydrogen damage
          mechanisms

I

Impact loads, 28, *see also* Toughness
Impact strength
  fracture mechanism, 29
  temperature effects, 29
Instantaneous zone, 61
  along direction of rolling, 93
  relative size, 61
  shape interpretation, 72
Intergranular corrosion, 109–110
  appearance, 109
  prevention, 110
Iron, *see* Carbon steels
IZ, *see* Instantaneous zone

K

Kletz, Trevor, 2

L

Laboratory for plant failure analysis
  equipment needed 22–23
Latent error roots, 2,8,12
Latino, Charles, 4,9
Liberty ship failures, 54
Lip seals, 175–176, 250–252